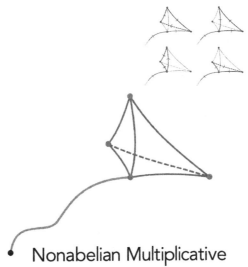

Nonabelian Multiplicative
Integration on Surfaces

Amnon Yekutieli

Ben Gurion University, Israel

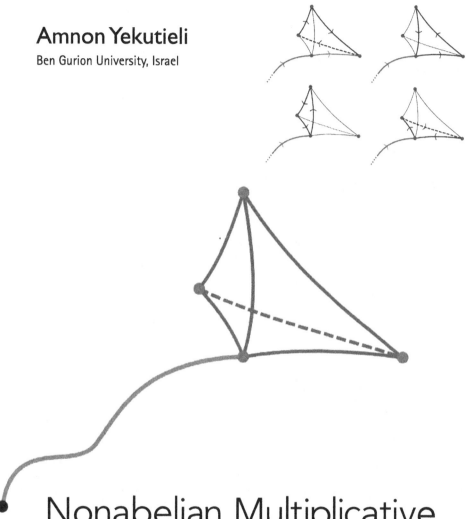

Nonabelian Multiplicative
Integration on Surfaces

World Scientific

NEW JERSEY · LONDON · SINGAPORE · BEIJING · SHANGHAI · HONG KONG · TAIPEI · CHENNAI · TOKYO

Published by

World Scientific Publishing Co. Pte. Ltd.

5 Toh Tuck Link, Singapore 596224

USA office: 27 Warren Street, Suite 401-402, Hackensack, NJ 07601

UK office: 57 Shelton Street, Covent Garden, London WC2H 9HE

Library of Congress Cataloging-in-Publication Data
Yekutieli, Amnon
    Nonabelian multiplicative integration on surfaces / by Amnon Yekutieli (Ben Gurion University, Israel).
        pages cm
    Includes bibliographical references and index.
    ISBN 978-9814663847 (hardcover : alk. paper)
    1. Noncommutative algebras. 2. Non-Abelian groups. 3. Surfaces, Algebraic. 4. Geometry, Algebraic. I. Title. II. Title: Non-abelian multiplicative integration on surfaces.
    QA251.4.Y45 2015
    512'.46--dc23
                                                2015024761

British Library Cataloguing-in-Publication Data
A catalogue record for this book is available from the British Library.

Printed in Singapore by Mainland Press Pte Ltd.

To FCR

For making this work possible...

Preface

Nonabelian multiplicative integration on curves is a classical theory, going back to Volterra in the 19th century. In differential geometry, this operation can be interpreted as the holonomy of a connection along a curve. In probability theory, this is a nonconstant continuous-time Markov process.

This book is about the 2-dimensional case. A rudimentary nonabelian multiplicative integration on surfaces was introduced in the 1920's by Schlesinger, but it is not widely known. Here I present a more sophisticated construction. The setup is a *Lie crossed module*: there is a Lie group H, together with an action on it by another Lie group G. The multiplicative integral is an element of H, and it is the limit of Riemann products. Each Riemann product involves a fractal decomposition of the surface into *kites* (triangles with strings connecting them to the base point). There is a twisting of the integrand, that comes from a 1-dimensional multiplicative integral along the strings, with values in the group G.

The main result of this work is the *3-dimensional nonabelian Stokes Theorem*. This result is new; only a special case of it was predicted (without proof) in papers in mathematical physics (by M. Kontsevich and others). My constructions and proofs are of a straightforward geometric and analytic nature. There are plenty of illustrations to clarify the geometric constructions (and color versions of these figures are available online, see web link below).

The motivation for my work was a problem in twisted deformation quantization of algebraic varieties (proposed by Kontsevich, and solved by me in a series of papers). This problem is related to algebraic geometry (the structure of gerbes), algebraic topology (nerves of 2-groupoids), and mathematical physics (nonabelian gauge theory). These relations are explained, and references to relevant papers are provided.

I am an expert in algebraic geometry, derived categories and deformation quantization. In this book, I take a journey into another area of mathematics – differential geometry – in which I am a novice. The outcome of this journey is a collection of constructions and theorems, obtained due to a fresh point of view. I hope this work will motivate and help other researchers interested in this subject. (Indeed, at least one such interaction has already materialized, in the paper on nerves of 2-groupoids by Tsygan et al. that is referenced.) The price, so to speak, of this kind of journey is that some passages in the book might be considered tedious or unduly detailed. For this, I apologize.

The web link to the online color figures is

http://www.worldscientific.com/worldscibooks/10.1142/9537

Amnon Yekutieli
April 2015

Acknowledgments

Work on this book began together with Fredrick Leitner, and I wish to thank him for his contributions, without which the book could not have been written. Thanks also to Michael Artin, Maxim Kontsevich, Damien Calaque, Lawrence Breen, Amos Nevo, Yair Glasner, Barak Weiss and Victor Vinnikov for assistance on various aspects of the book.

Contents

0 Introduction

0.1

In this book we establish a theory of *twisted nonabelian multiplicative integration of 2-forms on surfaces*.

Let us begin the exposition with a discussion of 1-*dimensional nonabelian multiplicative integration*. This goes back to the work of Volterra in the 19th century, and is also known by the names "path ordered exponential integration" and "holonomy of a connection along a path". See the book [DF] and the paper [KMR] for its history and various properties.

In our setup the 1-dimensional multiplicative integral looks like this. Let X be an n-dimensional manifold (by which we mean a differentiable manifold with corners), and let G be a Lie group with Lie algebra $\mathfrak{g}$, all over the field $\mathbb{R}$. We denote by

$$\Omega(X) = \bigoplus_{p=0}^{n} \Omega^p(X)$$

the de Rham algebra of X (i.e. the algebra of smooth differential forms). By smooth path in X we mean a smooth map $\sigma : \Delta^1 \to X$, where Δ^1 is the 1-dimensional simplex. Let

$$\alpha \in \Omega^1(X) \otimes \mathfrak{g}.$$

The *multiplicative integral of α on σ* is an element

(0.1.1) $\mathrm{MI}(\alpha \,|\, \sigma) \in G,$

obtained as the limit of *Riemann products*. This operation is re-worked and extended in Chapter 3 of our book.

0.2

For reasons explained in Section 0.11 of the Introduction, we found it necessary to devise a theory of 2-*dimensional nonabelian multiplicative integration*. Our work was guided by the problem at hand, plus ideas borrowed from the papers [BM, BS, Ko].

Instead of a single Lie group, the 2-dimensional operation involves a pair (G, H) of Lie groups, with a certain interaction between them. This

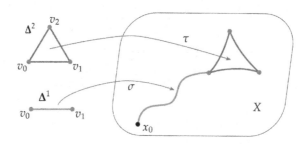

FIGURE 1. A smooth triangular kite (σ, τ) in the pointed manifold (X, x_0).

structure is called a *Lie crossed module*. A Lie crossed module

$$(0.2.1) \qquad\qquad \mathbf{C} = (G, H, \Psi, \Phi)$$

consists, in addition to the Lie groups H and G, of an analytic action Ψ of G on H by automorphisms of Lie groups, which we call the *twisting*; and of a map of Lie groups $\Phi : H \to G$, called the *feedback*. The conditions are that Φ is G-equivariant (with respect to Ψ and to the conjugation action Ad_G of G on itself); and

$$(0.2.2) \qquad\qquad \Psi \circ \Phi = \mathrm{Ad}_H .$$

See [BM, BS]. The integrand is now a pair (α, β), with

$$(0.2.3) \qquad\qquad \alpha \in \Omega^1(X) \otimes \mathfrak{g} \ \text{ and } \ \beta \in \Omega^2(X) \otimes \mathfrak{h}.$$

Here $\mathfrak{h}$ is the Lie algebra of H.

    Let x_0 be a point in X, so the pair (X, x_0) is a pointed manifold. The geometric data (the cycle) for the multiplicative integration is a *kite* (σ, τ) in (X, x_0). Let us denote by $v_0, \ldots, v_p$ the vertices of the p-dimensional simplex $\mathbf{\Delta}^p$. By definition a *smooth triangular kite* (σ, τ) in (X, x_0) consists of smooth maps $\sigma : \mathbf{\Delta}^1 \to X$ and $\tau : \mathbf{\Delta}^2 \to X$, such that $\sigma(v_0) = x_0$ and $\sigma(v_1) = \tau(v_0)$. See Figure 1 for an illustration.

    The next theorem summarizes our construction. It encapsulates many results scattered throughout the book.

Theorem 0.2.4 (Existence of MI on Triangles). *Let* $\mathbf{C} = (G, H, \Psi, \Phi)$ *be a Lie crossed module,* (X, x_0) *a pointed manifold, and* (α, β) *a pair of differential forms as in* (0.2.3). *Given any smooth triangular kite* (σ, τ) *in* (X, x_0), *there is an element*

$$\mathrm{MI}(\alpha, \beta \,|\, \sigma, \tau) \in H$$

called the twisted multiplicative integral of (α, β) on (σ, τ).

The operation $\mathrm{MI}(-, -)$ *enjoys these properties:*

(a) *The element* $\mathrm{MI}(\alpha, \beta \mid \sigma, \tau)$ *has an explicit formula as the limit of Riemann products.*

(b) *The operation* $\mathrm{MI}(-, -)$ *is functorial in* **C** *and* (X, x_0).

(c) *If H is abelian and G is trivial, then*

$$\mathrm{MI}(\alpha, \beta \mid \sigma, \tau) = \exp_H \left(\int_\tau \beta \right).$$

More on the construction in Sections 0.6-0.8 and 0.10 of the Introduction.

As far as we know, this is the first construction of a nonabelian multiplicative integration on surfaces of such generality. The very special case $G = H = \mathrm{GL}_m(\mathbb{R})$ was done by Schlesinger in the 1920's; cf. [DF, KMR].

0.3

Actually, in the body of the book we work in a much more complicated situation. Instead of a Lie crossed module (0.2.1), we work with a *Lie quasi crossed module with additive feedback* (see Chapter 5). The reason for the more complicated setup is that this is what was required, at the time, for our work on *twisted deformation quantization* in the paper [Ye3]. More on this in Section 0.11 below. In the Introduction we stick to the simpler setup of a Lie crossed module, which is interesting enough. Note however that all the results mentioned in the Introduction are valid also in the more complicated setup.

0.4

In this section we explain the nonabelian 2-dimensional Stokes Theorem. For this to hold it is necessary to impose a condition on the integrand (α, β). Recall the feedback $\Phi : H \to G$. Consider the induced Lie algebra homomorphism

$$\mathrm{Lie}(\Phi) : \mathfrak{h} \to \mathfrak{g}.$$

By tensoring this induces a homomorphism of differential graded Lie algebras

$$\phi : \Omega(X) \otimes \mathfrak{h} \to \Omega(X) \otimes \mathfrak{g}.$$

We say that (α, β) is a *connection-curvature pair* for **C**$/X$ if

(0.4.1) $$\phi(\beta) = \mathrm{d}(\alpha) + \tfrac{1}{2}[\alpha, \alpha]$$

in $\Omega^2(X) \otimes \mathfrak{g}$. (In [BM] this condition is called the *vanishing of the fake curvature*.)

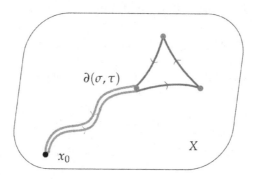

FIGURE 2. The boundary $\partial(\sigma,\tau)$ of the kite (σ,τ) from Figure 1.

The boundary of a triangular kite (σ,τ) is the closed path $\partial(\sigma,\tau)$ depicted in Figure 2.

Theorem 0.4.2 (Stokes Theorem for Triangles). *Let* $\mathbf{C} = (G, H, \Psi, \Phi)$ *be a Lie crossed module,* (X, x_0) *a pointed manifold, and* (α, β) *a connection-curvature pair for* $\mathbf{C}/X$. *Given any smooth triangular kite* (σ,τ) *in* (X, x_0), *one has*

$$\Phi\big(\mathrm{MI}(\alpha, \beta \,|\, \sigma, \tau)\big) = \mathrm{MI}(\alpha \,|\, \partial(\sigma, \tau))$$

in G.

The special case $G = H = \mathrm{GL}_m(\mathbb{R})$ is Schlesinger's Theorem (see [DF, KMR]).

0.5

We now approach the main result of the book, namely the nonabelian 3-dimensional Stokes Theorem.

A *smooth triangular balloon* in (X, x_0) is a pair (σ, τ), consisting of smooth maps $\sigma : \Delta^1 \to X$ and $\tau : \Delta^3 \to X$, such that $\sigma(v_0) = x_0$ and $\sigma(v_1) = \tau(v_0)$. See Figure 3 for an illustration.

The *boundary* of a balloon (σ, τ) is a sequence

$$\partial(\sigma, \tau) = \big(\partial_1(\sigma, \tau), \partial_2(\sigma, \tau), \partial_3(\sigma, \tau), \partial_4(\sigma, \tau)\big)$$

of triangular kites. See Figure 4. We write

(0.5.1) $$\mathrm{MI}(\alpha, \beta \,|\, \partial(\sigma, \tau)) := \prod_{1=1}^{4} \mathrm{MI}(\alpha, \beta \,|\, \partial_i(\sigma, \tau)),$$

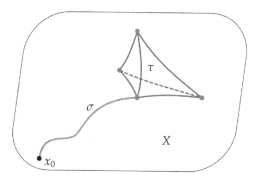

FIGURE 3. A smooth triangular balloon (σ, τ) in the pointed manifold (X, x_0).

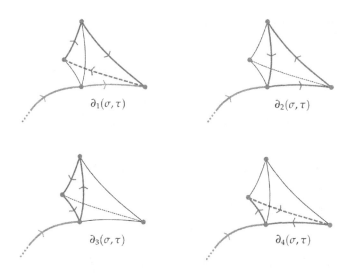

$\partial_1(\sigma, \tau)$ $\partial_2(\sigma, \tau)$

$\partial_3(\sigma, \tau)$ $\partial_4(\sigma, \tau)$

FIGURE 4. The four triangular kites that make up the sequence $\partial(\sigma, \tau)$. Here (σ, τ) is the triangular balloon from Figure 3.

where the order of the product is left to right.

Let $H_0 := \mathrm{Ker}(\Phi)$, which is a closed Lie subgroup of H. We call it the *inertia subgroup*, and $\mathfrak{h}_0 := \mathrm{Lie}(H_0)$ is called the *inertia subalgebra*. Note that H_0 is contained in the center of H, so H_0 is an abelian Lie group. A form

$$\gamma \in \Omega^p(X) \otimes \mathfrak{h}_0$$

is called an *inert p-form*.

We say that a form $\alpha \in \Omega^1(X) \otimes \mathfrak{g}$ is *tame connection* if it belongs to some connection-curvature pair (α, β). We show (in Section 8.3) that given a tame connection α and an inert 3-form γ, there is a well defined *twisted abelian multiplicative integral*

$$\mathrm{MI}(\alpha, \gamma \mid \sigma, \tau) \in H_0.$$

For $g \in G$ there is a Lie group automorphism $\Psi(g) : H \to H$. Its derivative is

$$\Psi_{\mathfrak{h}}(g) := \mathrm{Lie}(\Psi(g)) : \mathfrak{h} \to \mathfrak{h}.$$

As g varies we get a map of Lie groups

(0.5.2) $$\Psi_{\mathfrak{h}} : G \to \mathrm{GL}(\mathfrak{h}).$$

The derivative of $\Psi_{\mathfrak{h}}$ is a map of Lie algebras

$$\mathrm{Lie}(\Psi_{\mathfrak{h}}) : \mathfrak{g} \to \mathfrak{gl}(\mathfrak{h}) = \mathrm{End}(\mathfrak{h}).$$

This extends by tensor product to a homomorphism of differential graded Lie algebras

$$\psi_{\mathfrak{h}} : \Omega(X) \otimes \mathfrak{g} \to \Omega(X) \otimes \mathrm{End}(\mathfrak{h}).$$

Given a connection-curvature (α, β), its *3-curvature* is the form

(0.5.3) $$\gamma := \mathrm{d}(\beta) + \psi_{\mathfrak{h}}(\alpha)(\beta) \in \Omega^3 \otimes \mathfrak{h}.$$

(This is the name given in [BM].)

We can now state the main result of this book. (See Theorem 9.3.5 for the full version.)

Theorem 0.5.4 (Stokes Theorem for Tetrahedra). *Let* $\mathbf{C} = (G, H, \Psi, \Phi)$ *be a Lie crossed module,* (X, x_0) *a pointed manifold, and* (α, β) *a connection-curvature pair for* $\mathbf{C}/X$. *Let* γ *be the 3-curvature of* (α, β). *Then:*

(1) *The form* γ *is inert.*

(2) *Given any smooth triangular balloon* (σ, τ) *in* (X, x_0), *one has*

$$\mathrm{MI}(\alpha, \beta \mid \partial(\sigma, \tau)) = \mathrm{MI}(\alpha, \gamma \mid \sigma, \tau)$$

in H.

In this generality, Theorem 0.5.4 appears to have no precursor in the literature. The special (but most important) case, namely $\gamma = 0$, was taken to be true in some papers (e.g. [Ko, BS]), but no proofs were provided there.

Assertion (1) in the theorem can be viewed as a *generalized Bianchi identity*.

0.6

In this and the next few sections we will explain some of the work leading to the theorems mentioned above.

The methods used in the constructions and proofs are of two kinds: geometric and infinitesimal. The geometric methods consist of dividing kites into smaller ones, and studying the effect on the corresponding approximations (Riemann products). The infinitesimal methods involve Taylor expansions and estimates for the nonabelian exponential map.

Throughout most of the book we work with *linear quadrangular kites* in *polyhedra*, and with *piecewise smooth differential forms*, rather than with smooth triangular kites in manifolds and smooth differential forms. The reason for choosing quadrangular kites is mainly that it is much easier to carry out calculations of Taylor expansions on squares, as compared to triangles. Also the combinatorics of quadrangular kites and their binary subdivisions is simpler than that of triangles. See Sections 4.1-4.2. A review of the piecewise linear geometry of poyhedra, and of piecewise smooth differential forms, can be found in Chapter 1.

The key technical result that allows us to calculate nonabelian products is Theorem 2.1.2 on estimates for the nonabelian exponential map. These estimates are obtained from the CBH formula. Chapter 2 is devoted to the proof of these estimates.

For heuristic purposes we introduce the concept of "tiny scale". By tiny scale (depending on context of course) we mean geometric or algebraic elements that are so small that the relevant estimates (arising from the CBH formula or Taylor expansion) apply to them. See Remark 3.3.5.

0.7

For $n \geq 0$ let $\mathbf{I}^n$ be the n-dimensional cube. We give names to some vertices: $v_0 := (0, \ldots, 0)$, and

$$v_i := (0, \ldots, 1, \ldots, 0)$$

with 1 in the i-th position, for $i = 1, \ldots, n$. The base point for $\mathbf{I}^n$ is v_0.

Let (X, x_0) be a pointed polyhedron. By *linear string* in X we mean a sequence $\sigma = (\sigma_1, \ldots, \sigma_m)$ of piecewise linear maps $\sigma_i : \mathbf{I}^1 \to X$, such that $\sigma_i(v_1) = \sigma_{i+1}(v_0)$. The maps σ_i are called the pieces of σ. In Chapter 3 we construct the nonabelian multiplicative integral $\mathrm{MI}(\alpha \,|\, \sigma)$ of a piecewise smooth differential form $\alpha \in \Omega^1_{\mathrm{pws}}(X) \otimes \mathfrak{g}$ on a string σ. This construction is quite simple and essentially the same as the classical one. We also prove a few basic properties of this MI.

0.8

In Chapter 4 we construct the 2-dimensional nonabelian multiplicative integral $\mathrm{MI}(\alpha, \beta \mid \sigma, \tau)$. Here (σ, τ) is a linear quadrangular kite in the pointed polyhedron (X, x_0). By definition this means that σ is a string in X, and $\tau : \mathbf{I}^2 \to X$ is a linear map. The conditions are that the initial point of σ is x_0, and the terminal point of σ is $\tau(v_0)$. The integrand (α, β) consists of piecewise smooth differential forms:

$$\alpha \in \Omega^1_{\mathrm{pws}}(X) \otimes \mathfrak{g} \text{ and } \beta \in \Omega^2_{\mathrm{pws}}(X) \otimes \mathfrak{h}.$$

The coarse approximation of the 2-dimensional nonabelian multiplicative integral is as follows. Given a kite (σ, τ) in (X, x_0), let

$$g := \mathrm{MI}(\alpha \mid \sigma) \in G;$$

so applying the operator $\Psi_{\mathfrak{h}}(g)$ from (0.5.2) we have a new (twisted) 2-form

$$\Psi_{\mathfrak{h}}(g)(\beta) \in \Omega^2_{\mathrm{pws}}(X) \otimes \mathfrak{h}.$$

We define the *basic Riemann product* of (α, β) on (σ, τ) to be

$$(0.8.1) \qquad \mathrm{RP}_0(\alpha, \beta \mid \sigma, \tau) := \exp_H \left(\int_\tau \Psi_{\mathfrak{h}}(g)(\beta) \right) \in H.$$

(Actually in the body of the book we use another formula for $\mathrm{RP}_0(\alpha, \beta \mid \sigma, \tau)$, that converges faster; see Definition 4.3.3 and Remark 4.3.4.)

For the limiting process we introduce the *binary subdivisions* of $\mathbf{I}^2$. The k-th binary subdivision is the cellular decomposition of $\mathbf{I}^2$ into 4^k equal squares, and we denote it by $\mathrm{sd}^k \mathbf{I}^2$. The 1-skeleton of $\mathrm{sd}^k \mathbf{I}^2$ is denoted by $\mathrm{sk}_1 \mathrm{sd}^k \mathbf{I}^2$. Its fundamental group (based at v_0) is denoted by $\pi_1(\mathrm{sk}_1 \mathrm{sd}^k \mathbf{I}^2)$. It is very important that the group $\pi_1(\mathrm{sk}_1 \mathrm{sd}^k \mathbf{I}^2)$ is a free group on 4^k generators. We say that a kite (σ, τ) is *patterned on* $\mathrm{sd}^k \mathbf{I}^2$ if for every piece σ_i of the string σ, the image $\sigma_i(\mathbf{I}^1)$ is a 1-cell of $\mathrm{sd}^k \mathbf{I}^2$, and $\tau(\mathbf{I}^2)$ is a 2-cell of $\mathrm{sd}^k \mathbf{I}^2$.

A *tessellation* of $\mathbf{I}^2$ is by definition a sequence $\big((\sigma_i, \tau_i)\big)_{i=1,\dots,4^k}$ of square kites in $(\mathbf{I}^2, v_0)$, each patterned on $\mathrm{sd}^k \mathbf{I}^2$, satisfying the following topological condition. Let us denote by $[\partial(\sigma_i, \tau_i)]$ the element of $\pi_1(\mathrm{sk}_1 \mathrm{sd}^k \mathbf{I}^2)$

represented by the closed string $\partial(\sigma_i, \tau_i)$. Then

$$(0.8.2) \qquad \prod_{1=1}^{4^k} [\partial(\sigma_i, \tau_i)] = [\partial \mathbf{I}^2]$$

in $\pi_1(\mathrm{sk}_1 \, \mathrm{sd}^k \, \mathbf{I}^2)$.

For the construction we choose a particular tessellation for every k. It is called the *k-th binary tessellation*, and the notation is $\mathrm{tes}^k \, \mathbf{I}^2$. The actual definition of $\mathrm{tes}^k \, \mathbf{I}^2$ is not important (since it is quite arbitrary, and chosen for convenience). All that is important are its two properties:

- It is a tessellation; i.e. equation (0.8.2) holds.
- It has a recursive (self similar) nature.

For fixed k and a kite (σ, τ) we obtain, by a simple geometric operation, the k-binary tessellation

$$\mathrm{tes}^k(\sigma, \tau) = \left(\mathrm{tes}_i^k(\sigma, \tau)\right)_{i=1,\ldots,4^k}$$

of (σ, τ), which is a sequence of kites in (X, x_0). (See Definition 4.2.7.)

For every $i \in \{1, \ldots, 4^k\}$ we have the basic Riemann product

$$\mathrm{RP}_0(\alpha, \beta \mid \mathrm{tes}_i^k(\sigma, \tau))$$

on the kite $\mathrm{tes}_i^k(\sigma, \tau)$. We define the *k-th Riemann product* to be

$$(0.8.3) \qquad \mathrm{RP}_k(\alpha, \beta \mid \sigma, \tau) := \prod_{1=1}^{4^k} \mathrm{RP}_0(\alpha, \beta \mid \mathrm{tes}_i^k(\sigma, \tau)).$$

In Theorem 4.4.15 we prove that the limit

$$\lim_{k \to \infty} \mathrm{RP}_k(\alpha, \beta \mid \sigma, \tau)$$

exists in H. The proof goes like this: for sufficiently large k the kites $\mathrm{tes}_i^k(\sigma, \tau)$ are all tiny. We use estimates to show that the limits

$$\lim_{k' \to \infty} \mathrm{RP}_{k'}(\alpha, \beta \mid \mathrm{tes}_i^k(\sigma, \tau))$$

exist for every i. Due to the recursive nature of the binary tessellations, this is enough. We can finally define

$$(0.8.4) \qquad \mathrm{MI}(\alpha, \beta \mid \sigma, \tau) := \lim_{k \to \infty} \mathrm{RP}_k(\alpha, \beta \mid \sigma, \tau).$$

0.9

In Chapter 6 we prove the 2-dimensional Stokes Theorem. It is the same as Theorem 0.4.2, except that it talks about piecewise smooth forms on a pointed polyhedron and linear quadrangular kites. Again the proof is by reduction (using the recursive nature of the binary tessellations) to the case of a tiny kite. And then we use approximations (both of Taylor expansions and the exponential map) to do the calculation.

A very important technical consequence of the 2-dimensional Stokes Theorem is:

Theorem 0.9.1 (Fundamental Relation). *Let* $\mathbf{C} = (G, H, \Psi, \Phi)$ *be a Lie crossed module,* (X, x_0) *a pointed polyhedron,* (α, β) *a piecewise smooth connection-curvature pair for* $\mathbf{C}/X$, *and* (σ, τ) *a linear quadrangular kite in* (X, x_0). *Consider the elements*

$$g := \mathrm{MI}\big(\alpha \,|\, \partial(\sigma, \tau)\big) \in G$$

and

$$h := \mathrm{MI}(\alpha, \beta \,|\, \sigma, \tau) \in H.$$

Then

$$\Psi(g) = \mathrm{Ad}_H(h)$$

as automorphisms of the Lie group H.

In Chapter 8 we prove the first version of the main result of the book, namely the 3-dimensional Stokes Theorem (Theorem 8.6.6). This is like Theorem 0.5.4, only for piecewise smooth connection-curvature pairs, and for linear quadrangular balloons.

The strategy of the proof is this. Using Theorem 0.9.1, and a lot of combinatorics of free groups (done in Chapter 7), we reduce the problem to the case of a tiny balloon. And for a tiny balloon we use Taylor expansions and the estimates from Theorem 2.1.2.

0.10

Finally, in Chapter 9, we show how to pass from quadrangular kites to triangular ones. This is very easy, using the following trick. The triangle Δ^2 is naturally embedded in the square $\mathbf{I}^2$; and there is a piecewise linear retraction $h : \mathbf{I}^2 \to \Delta^2$. Consider the "universal triangular kite" (σ', τ'), and the "universal quadrangular kite" (σ', τ''), both in $(\mathbf{I}^2, v_0)$; see Figures 34, 35 and 36. There is a piecewise linear retraction

$$g : \mathbf{I}^2 \to \sigma'(\mathbf{I}^1) \cup \tau'(\mathbf{I}^2).$$

These maps are related by

$$\tau' \circ h = g \circ \tau'',$$

as maps $\mathbf{I}^2 \to \mathbf{I}^2$.

Given a smooth kite (σ, τ) in a pointed manifold (X, x_0), there a piecewise smooth map $f : \mathbf{I}^2 \to X$ such that $f \circ \sigma' = \sigma$ and $f \circ \tau' = \tau$. Given a smooth connection-curvature pair (α, β) for $\mathbf{C}/X$, the pair

$$(\alpha', \beta') := (f^*(\alpha), f^*(\beta))$$

is a piecewise smooth connection-curvature pair (α, β) for $\mathbf{C}/\mathbf{I}^2$. We define

(0.10.1) $\mathrm{MI}(\alpha, \beta \,|\, \sigma, \tau) := \mathrm{MI}(\alpha', \beta' \,|\, \sigma', \tau'').$

The results proved for quadrangular kites can be easily transferred to triangular kites by similar geometric tricks.

0.11

The reason we became interested in nonabelian multiplicative integration is its application to twisted deformation quantization of algebraic varieties, as developed in our paper [Ye3] (see also the survey article [Ye4]). Twisted deformations of an algebraic variety X are very close to gerbes and to stacks of algebroids (in the sense of [Ko]). Our original strategy was to classify these twisted deformations in terms of Maurer-Cartan solutions in a DG Lie algebra obtained as the commutative Čech resolution of a sheaf of DG Lie algebras on X. The multiplicative integration was supposed to be a key component of this construction. Some of the details are explained in Sections 5.6, 5.7 and 9.5.

In the meanwhile we discovered a more direct approach to the classification of twisted deformations, using descent data in cosimplicial crossed groupoids, and a related invariance result (see Sections 5-6 in the latest version of [Ye3], and Theorem 0.1 in [Ye6]).

In the version of [Ye3] dated 26 August 2009 there was a "result" labeled Theorem 11.2, about cosimplicial quantum type DG Lie algebras, Deligne crossed groupoids and descent data. The proof of this "result" was only sketched, and we never completed the proof (it was supposed to have appeared in a separate paper). It is now presented as Conjecture 9.5.2 at the end of this book.

Speaking of conjectures, we should mention Conjecture 9.4.1 regarding the rationality of the 2-dimensional multiplicative integral when the Lie groups G and H are unipotent algebraic groups.

Despite the fact that the original motivation for this book no longer exists, we believe that the constructions and results presented here have independent value. This can be seen for instance by the role our work has in the recent paper [BGNT]. See the next section for a discussion of [BGNT] and other related papers.

0.12

We end the introduction with a discussion of recent related work. Our 2-dimensional nonabelian multiplicative integral is closely related to the *differential geometry of gerbes*, as developed by Breen and Messing [BM]. Indeed, the notion of fake curvature, that goes into our definition of connection-curvature pair, and the notion of 3-curvature, are both taken from [BM].

There is a series of papers by Baez, Schreiber and Waldorf ([BS], [SW1], [SW2] and others) on *nonabelian gauge theory*. They develop a theory of nonabelian curvature that seems to be very similar to our 2-dimensional nonabelian multiplicative integral, even though the flavor of their work is very different (it is much more abstract). It seems that [SW2] contains a proof of the 3-dimensional Stokes Theorem in the special case where the 3-curvature is zero.

Very recently the paper [BGNT] by Bressler, Gorokhovsky, Nest and Tsygan came out. The goal of this paper is to prove that several existing notions of the nerve of a 2-groupoid are equivalent (as simplicial sets). The comparison between the Hinich simplicial set and the Deligne-Getzler simplicial set is done using 2-dimensional nonabelian multiplicative integration. The method seems to be quite similar, at least morally, to our original program (see previous section and Conjecture 9.5.2). The 2-dimensional nonabelian multiplicative integral that is developed in Sections 4-5 of [BGNT] is inspired by our work, and they obtain a version of our 3-dimensional Stokes Theorem in their setting (see [BGNT, Theorems 5.1 and 5.2]).

The difference between the construction in [BGNT] and ours is that they work in the special case of a Lie crossed module (G, H, Ψ, Φ) consisting of unipotent algebraic groups G and H, and an algebraic connection-curvature pair (α, β) with vanishing 3-curvature. The construction goes like this: they write a particular partial differential equation, and the multiplicative integral is its unique solution. This construction is significantly shorter than ours; and presumably it coincides with our construction in this setup. Cf. Section 5.5 of our book. See the end of Section 9.4 regarding the relation between the work of [BGNT] and Conjecture 9.4.1.

1 Polyhedra and Piecewise Smooth Geometry

1.1 Conventions

In this book we work over the field $\mathbb{R}$ of real numbers. We denote by $\mathbf{A}^n = \mathbf{A}^n(\mathbb{R})$ the real n-dimensional affine space, with the usual smooth (i.e. C^∞ differentiable) manifold structure, and the standard euclidean metric. The symbol $\otimes$ stands for $\otimes_{\mathbb{R}}$. The coordinate functions on $\mathbf{A}^n$ are $t_1, \ldots, t_n$.

1.2 Embedded Polyhedra

By linear map $f : \mathbf{A}^m \to \mathbf{A}^n$ we mean the composition of a linear homomorphism and a translation. Thus the group of invertible linear maps of $\mathbf{A}^n$ is $\mathrm{GL}_n(\mathbb{R}) \ltimes \mathbb{R}^n$. By linear subset X of $\mathbf{A}^n$ we mean the zero locus of some set of linear functions (not necessarily homogeneous). In other words X is the image of some injective linear map $f : \mathbf{A}^m \to \mathbf{A}^n$; and then we let m be the dimension of X.

Given a set $S \subset \mathbf{A}^n$, its linear (resp. convex) hull is the smallest linear (resp. convex) subset of $\mathbf{A}^n$ containing S. In case S is finite, say $S = \{x_1, \ldots, x_m\}$, then a point x is in the linear (resp. convex) hull of S if and only if $x = \sum_{i=1}^m a_i x_i$ for some real numbers $a_1, \ldots, a_m$ satisfying $\sum a_i = 1$ (resp. and $a_i \geq 0$).

By *embedded polyhedron* we always mean a convex bounded finite polyhedron in $\mathbf{A}^n$, for some n. Namely a polyhedron X in $\mathbf{A}^n$ is the convex hull of a finite subset of $\mathbf{A}^n$. A point $x \in X$ is called a *vertex* if it is not in the convex hull of any finite subset of $X - \{x\}$. If we denote by S the set of vertices of X, then S is finite, and X is the convex hull of S. The dimension $\dim X$ is the dimension of the linear hull of X, and we call $\mathbf{A}^n$ the *ambient linear space* of X. The standard euclidean metric on $\mathbf{A}^n$ restricts to a global metric (a distance function) dist_X on X. We denote by $\mathrm{diam}(X)$ the diameter of X, namely the maximal distance between any two points. The linear functions on $\mathbf{A}^n$ (spanned by $t_1, \ldots, t_n$) restrict to functions $X \to \mathbb{R}$, that we also call linear functions. The $\mathbb{R}$-module

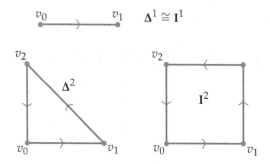

FIGURE 5. The polyhedra $\Delta^1 \cong I^1$, Δ^2 and I^2, with their vertices and orientations.

$\mathcal{O}_{\mathrm{lin}}(X)$ of linear functions has rank equal to $m := \dim X$. An $\mathbb{R}$-basis $\{s_i\}_{i=1,\ldots,m}$ of $\mathcal{O}_{\mathrm{lin}}(X)$ is called a *linear coordinate system* on X.

We shall mostly encounter two kinds of embedded polyhedra. The first is the *n-dimensional cube* I^n. This is the subset of $\mathbf{A}^n$ defined by the inequalities $0 \le t_1, \ldots, t_n \le 1$. As a convex set it is spanned by its 2^n vertices. See Figure 5 for an illustration.

The second kind of embedded polyhedron is the *n-dimensional simplex* Δ^n, embedded in $\mathbf{A}^{n+1}$. We use barycentric coordinates $t_0, t_1, \ldots, t_n$ on $\mathbf{A}^{n+1}$ when dealing with the simplex. In these coordinates Δ^n is the subset of $\mathbf{A}^{n+1}$ defined by the conditions $0 \le t_0, \ldots, t_n$ and $\sum_{i=0}^n t_i = 1$. As a convex set it is spanned by its $n+1$ vertices.

Suppose X and Y are both embedded polyhedra. A *linear map* $f : X \to Y$ is a function that extends to a linear map between the ambient linear spaces. Such a map f is determined by its value on a sequence $(v_0, \ldots, v_n)$ of linearly independent vertices of X, where $n := \dim X$. Thus we shall often describe the linear map f as

$$f(v_0, \ldots, v_n) = (w_0, \ldots, w_n),$$

or just $f = (w_0, \ldots, w_n)$, where $w_i := f(v_i)$.

In case X is the cube I^n, embedded in $\mathbf{A}^n$, we shall look at the following linearly independent vertices: v_0 is the origin $(0, \ldots, 0)$, and for $i = 1, \ldots, n$ we take

(1.2.1) $v_i := (0, \ldots, 1, \ldots, 0),$

with 1 in the i-th position. In case of the simplex Δ^n, embedded in $\mathbf{A}^{n+1}$ with barycentric coordinates, we look at all its vertices, and use the notation $(v_0, \ldots, v_n)$, where v_i is as in equation (1.2.1).

1.3 Smooth Maps and Forms

The algebra of smooth real functions on $\mathbf{A}^n$ is denoted by $\mathcal{O}(\mathbf{A}^n)$. The de Rham DG (differential graded) algebra of smooth differential forms on $\mathbf{A}^n$ is

$$\Omega(\mathbf{A}^n) = \bigoplus_{p=0}^{n} \Omega^p(\mathbf{A}^n),$$

with differential d and wedge product $\wedge$.

Let X be an m-dimensional polyhedron embedded in $\mathbf{A}^n$. The set X is a compact topological space, its interior $\mathrm{Int}(X)$ is an m-dimensional manifold, and the boundary ∂X is a finite union of $(m-1)$-dimensional polyhedra. But X also has a differentiable structure.

Definition 1.3.1. Let $X \subset \mathbf{A}^n$ be an embedded polyhedron, and let $U \subset X$ be an open set.

(1) A function $f : U \to \mathbb{R}$ is said to be *smooth* if it extends to a smooth function $\tilde{f} : \tilde{U} \to \mathbb{R}$ on some open neighborhood $\tilde{U}$ of U in $\mathbf{A}^n$.

(2) A map $f : U \to \mathbf{A}^p$ is called smooth if the functions $t_i \circ f : U \to \mathbb{R}$ are smooth for all $i \in \{1, \ldots, p\}$.

(3) Suppose $Y \subset \mathbf{A}^p$ is an embedded polyhedron, and $V \subset Y$ is an open set. A map $f : U \to V$ is called smooth if the composed map $f : U \to \mathbf{A}^p$ is smooth.

(4) Let V be an open subset of an embedded polyhedron. A map $f : U \to V$ is called a *diffeomorphism* of it is bijective, and both f and f^{-1} are smooth.

We denote by $\mathcal{O}(U)$ the $\mathbb{R}$-algebra of all smooth functions $U \to \mathbb{R}$. A smooth map $f : U \to V$ between open subsets of embedded polyhedra induces an $\mathbb{R}$-algebra homomorphism $f^* : \mathcal{O}(V) \to \mathcal{O}(U)$, namely pullback.

Suppose X is an embedded polyhedron and $U \subset X$ is an open set. We let $\mathrm{Int}(U) := U \cap \mathrm{Int}(X)$ and $\partial(U) := U \cap \partial(X)$. Note that $\mathrm{Int}(U)$ is a manifold, and it is dense in U.

Definition 1.3.2. Let $X \subset \mathbf{A}^n$ be an embedded polyhedron, and let $U \subset X$ be an open set. A *smooth differential p-form* on U is a differential form $\alpha \in \Omega^p(\mathrm{Int}(U))$, that extends to a form $\tilde{\alpha} \in \Omega^p(\tilde{U})$ on some open neighborhood $\tilde{U}$ of U in $\mathbf{A}^n$.

We denote by $\Omega^p(U)$ the set of differential p-forms on U, and we let

$$\Omega(U) := \bigoplus_{p \geq 0} \Omega^p(U).$$

This is a DG subalgebra of $\Omega(\mathrm{Int}(U))$.

Suppose U and V are open subsets of embedded polyhedra. A smooth map $f : U \to V$ induces a DG algebra homomorphism $f^* : \Omega(V) \to \Omega(U)$. If f is an inclusion then we usually write $\alpha|_U := f^*(\alpha)$ for $\alpha \in \Omega(V)$.

Let X be an m-dimensional embedded polyhedron, and $U \subset X$ a nonempty open set. Then $\Omega^p(U)$ is a free $\mathcal{O}(U)$-module of rank $\binom{m}{p}$. Indeed, if we choose a linear coordinate system $\{s_i\}_{i=1,\dots,m}$ on X, then the family

(1.3.3) $\{ds_{i_1} \wedge \cdots \wedge ds_{i_p}\}_{1 \le i_1 < \cdots < i_p \le m}$

is a basis for $\Omega^p(U)$ as $\mathcal{O}(U)$-module.

Definition 1.3.4. Let $s = \{s_i\}_{i=1,\dots,m}$ be a linear coordinate system on the m-dimensional polyhedron X. Given a form $\alpha \in \Omega^p(X)$, it has a unique expansion as a sum

$$\alpha = \sum_i \tilde{\alpha}_i \cdot ds_{i_1} \wedge \cdots \wedge ds_{i_p},$$

where $i = (i_1, \dots, i_p)$ runs over the set of strictly increasing multi-indices in the range $\{1, \dots, m\}$, and $\tilde{\alpha}_i \in \mathcal{O}(X)$. The functions $\tilde{\alpha}_i$ are called the *coefficients* of α with respect to the linear coordinate system s.

Let us denote by $\Omega_{\mathrm{const}}(\mathbf{A}^n)$ the $\mathbb{R}$-subalgebra of $\Omega(\mathbf{A}^n)$ generated by the elements $dt_1, \dots, dt_n$. Now suppose X is an m-dimensional polyhedron in $\mathbf{A}^n$. The image of $\Omega_{\mathrm{const}}(\mathbf{A}^n)$ under the canonical surjection $\Omega(\mathbf{A}^n) \to \Omega(X)$ is denoted by $\Omega_{\mathrm{const}}(X)$, and its elements are called *constant differential forms*. There is an isomorphism of graded algebras

$$\Omega(X) \cong \mathcal{O}(X) \otimes \Omega_{\mathrm{const}}(X).$$

If we choose a linear coordinate system $s = \{s_i\}_{i=1,\dots,m}$ on X, then (1.3.3) is an $\mathbb{R}$-basis for $\Omega^p_{\mathrm{const}}(X)$.

Definition 1.3.5. Given a smooth form $\alpha \in \Omega^p(X)$ and a point $x \in X$, we define the *constant form* associated to α at x to be

$$\alpha(x) := \sum_i \tilde{\alpha}_i(x) \cdot ds_{i_1} \wedge \cdots \wedge ds_{i_p} \in \Omega^p_{\mathrm{const}}(X),$$

where the functions $\tilde{\alpha}_i$ are the coefficients of α relative to some linear coordinate system $s = \{s_i\}$, as in Definition 1.3.4.

Note that $\alpha(x)$ is independent of the linear coordinate system.

Let $f : X \to Y$ be a linear map between embedded polyhedra. If g is a linear function on Y, then $f^*(g) = g \circ f$ is a linear function on X; and if $\alpha \in \Omega_{\mathrm{const}}(Y)$, then $f^*(\alpha) \in \Omega_{\mathrm{const}}(X)$.

1.4 Manifolds with Sharp Corners

Manifolds with corners are modeled on $\mathbb{R}^n_{\geq 0}$; cf. [Le] or [Jo]. However not all polyhedra are manifolds with corners; so we introduce a variant in which the local model is an embedded polyhedron. See Remark 1.4.4 below.

Definition 1.4.1. Let X be a topological space.

(1) A *smooth chart with sharp corners* on X is a pair (U, ϕ), where U is an open subset of an embedded polyhedron, and $\phi : U \to X$ is a map such that $\phi(U)$ is open in X and $\phi : U \to \phi(U)$ is a homeomorphism.

(2) Two smooth charts with sharp corners (U_1, ϕ_1) and (U_2, ϕ_2) are compatible if

$$\phi_2^{-1} \circ \phi_1 : \phi_1^{-1}(\phi_1(U_1) \cap \phi_2(U_2)) \to \phi_2^{-1}(\phi_1(U_1) \cap \phi_2(U_2))$$

is a diffeomorphism.

(3) A *smooth atlas with sharp corners* on X is a collection $\{(U_i, \phi_i)\}$ of compatible charts with sharp corners, such that $X = \bigcup_i \phi_i(U_i)$.

Definition 1.4.2. A *smooth manifold with sharp corners* is a Hausdorff second countable topological space X endowed with a smooth atlas with sharp corners $\{(U_i, \phi_i)\}$.

Definition 1.4.3. Let X and Y be smooth manifolds with sharp corners, with atlases $\{(U_i, \phi_i)\}$ and $\{(V_j, \psi_j)\}$ respectively. A map $f : X \to Y$ is called *smooth* if for every i, j the map

$$\psi_j^{-1} \circ f \circ \phi_i : (f \circ \phi_i)^{-1}(\psi_j(V_j)) \to V_j$$

is smooth.

Remark 1.4.4. If we were to restrict the class of polyhedra in Definition 1.4.2 to cubes (or simplices), then we would recover the usual definition of manifold with corners. "Sharp corners" occur when too many faces of a polyhedron meet at a vertex (to be precise, more than the dimension of the manifold; e.g. at the top vertex of a pyramid with square base).

There are several notions of smooth maps between manifolds with corners in the literature. Suppose X and Y are smooth manifolds with corners, and $f : X \to Y$ is a smooth map in the sense of Definition 1.4.3.

Then f is called "weakly smooth" in [Jo]. In order for f to be smooth according to [Jo] it must satisfy a condition involving the boundary strata of X and Y; such a condition cannot even be stated when sharp corners are present.

Proposition 1.4.5. *The category of smooth manifolds with sharp corners contains the following categories as full subcategories:*

(1) *The category of smooth manifolds (without boundary) and smooth maps between them.*

(2) *The category of embedded polyhedra and smooth maps between them (Definition 1.3.1(3)).*

We leave out the easy proof.

A function $f : X \to \mathbb{R}$ is called smooth if when viewed as a map $f : X \to \mathbf{A}^1$ it is smooth. We denote by $\mathcal{O}(X)$ the $\mathbb{R}$-algebra of smooth functions.

If U is an open subset of a smooth manifold with sharp corners X, then U inherits a structure of smooth manifold with sharp corners.

A smooth manifold with sharp corners X has an intrinsic decomposition $X = \partial X \coprod \text{Int}(X)$, where the interior $\text{Int}(X)$ is a manifold without boundary, and the boundary ∂X is a manifold with sharp corners. The condition for a point $x \in X$ to belong to $\text{Int}(X)$ is that it has an open neighborhood that is isomorphic, as manifold with sharp corners, to a manifold without boundary.

We say that a manifold with sharp corners X has dimension n if the manifold $\text{Int}(X)$ has dimension n. (Of course every connected component of X has a dimension.) In this case ∂X has dimension $n - 1$.

Let X be a smooth manifold with sharp corners with atlas $\{(U_i, \phi_i)\}$. For every i we have a diffeomorphism $\phi_i : \text{Int}(U_i) \to \text{Int}(\phi_i(U_i))$ between manifolds without boundaries, and $\text{Int}(\phi_i(U_i)) = \phi_i(U_i) \cap \text{Int}(X)$. The module of smooth differentials $\Omega^p(U_i)$ was introduced in Definition 1.3.2.

Definition 1.4.6. Let X be a smooth manifold with sharp corners with atlas $\{(U_i, \phi_i)\}$. A *smooth differential p-form* on X is a differential form $\alpha \in \Omega^p(\text{Int}(X))$, such that for every i the form

$$\phi_i^*(\alpha|_{\text{Int}(\phi_i(U_i))}) \in \Omega^p(\text{Int}(U_i))$$

extends to a form $\tilde{\alpha}_i \in \Omega^p(U_i)$.

We denote by $\Omega^p(X)$ the set of differential p-forms on X, and we let

$$\Omega(X) := \bigoplus_{p \geq 0} \Omega^p(X).$$

This is a DG subalgebra of $\Omega(\mathrm{Int}(X))$. Of course $\Omega^0(X) = \mathcal{O}(X)$.

A smooth map $f : X \to Y$ between manifolds with sharp corners induces a DG algebra homomorphism $f^* : \Omega(Y) \to \Omega(X)$.

Convention 1.4.7. For the rest of the book, by "manifold" we always mean a smooth manifold with sharp corners, as defined in Definition 1.4.2. Smooth maps between manifolds are in the sense of Definition 1.4.3.

1.5   Abstract Polyhedra

It turns out that the geometric structure of a polyhedron is easy to encode, and it is nicer to work with polyhedra that are not embedded.

By a *global metric* on a topological space Z we mean a distance function $\mathrm{dist}_Z : Z^2 \to \mathbb{R}_{\geq 0}$ that induces the topology of Z. Recall that if Z is an embedded polyhedron in $\mathbf{A}^m$, then Z inherits a global metric from the ambient affine space, and also an $\mathbb{R}$-module of linear functions $\mathcal{O}_{\mathrm{lin}}(Z) \subset \mathcal{O}(Z)$.

Definition 1.5.1. A *polyhedron* is a smooth manifold with sharp corners X, together with a global metric dist_X and an $\mathbb{R}$-submodule $\mathcal{O}_{\mathrm{lin}}(X) \subset \mathcal{O}(X)$. The condition is that there exists an isomorphism of smooth manifolds with sharp corners $f : X \xrightarrow{\simeq} Z$ for some embedded polyhedron $Z \subset \mathbf{A}^m$, with these two properties:

   (i) f is an isometry; i.e. $f : (X, \mathrm{dist}_X) \to (Z, \mathrm{dist}_Z)$ is an isomorphism of metric spaces.

  (ii) The $\mathbb{R}$-algebra isomorphism $f^* : \mathcal{O}(Z) \to \mathcal{O}(X)$ satisfies $f^*(\mathcal{O}_{\mathrm{lin}}(Z)) = \mathcal{O}_{\mathrm{lin}}(X)$.

The composed map $f : X \to \mathbf{A}^m$ is called a *linear metric embedding*. If only property (ii) holds, then $f : X \to \mathbf{A}^m$ is called a *linear embedding*.

In terms of this definition, an embedded polyhedron is a polyhedron X equipped with a particular linear metric embedding $f : X \to \mathbf{A}^m$.

Definition 1.5.2. Let X and Y be polyhedra, and let $f : X \to Y$ be a map.

  (1) We say f is *smooth* if it is a smooth map of smooth manifolds with sharp corners (see Definition 1.4.3).

  (2) We say f is *linear* if $f^*(\mathcal{O}_{\mathrm{lin}}(Y)) \subset \mathcal{O}_{\mathrm{lin}}(X)$.

By *linear coordinate system* on X we mean an $\mathbb{R}$-basis of $\mathcal{O}_{\mathrm{lin}}(X)$.

Proposition 1.5.3. *Let X be a polyhedron, and let $s = \{s_i\}_{i=1,\dots,n}$ be linear coordinate system on X. Consider the induced smooth map $f : X \to \mathbf{A}^n$. Then the image $Z := f(X)$ is an n-dimensional embedded polyhedron in $\mathbf{A}^n$,*

$f : X \to Z$ is an isomorphism of smooth manifolds with sharp corners, and $f^*(\mathcal{O}_{\mathrm{lin}}(Z)) = \mathcal{O}_{\mathrm{lin}}(X)$, i.e. the map $f : X \to \mathbf{A}^n$ is a linear embedding.

The proof is left as an exercise.

Suppose we are given linear embeddings $X \to \mathbf{A}^m$ and $Y \to \mathbf{A}^n$. It is easy to see that a map $f : X \to Y$ is linear in the sense of Definition 1.5.2 if and only if f extends to a linear map $\mathbf{A}^m \to \mathbf{A}^n$.

Note that we said nothing about the metrics in Proposition 1.5.3.

Definition 1.5.4. An *orthonormal linear coordinate system* on X is a linear coordinate system s such that the induced map $f : X \to \mathbf{A}^n$ is a linear metric embedding.

Let s be an orthonormal linear coordinate system on X. Then for any $x, y \in X$ the distance between them is

$$\mathrm{dist}_X(x, y) = \left(\sum_{i=1}^n \left(s_i(x) - s_i(y) \right)^2 \right)^{1/2}.$$

We can view our polyhedra as "Riemannian manifolds", in the following sense. First note that we can talk about the DG algebra $\Omega_{\mathrm{const}}(X)$ of constant forms: it is the DG algebra generated over $\mathbb{R}$ by the family of forms $\{\mathrm{d}s_i\}_{i=1,\ldots,n}$, where $s = \{s_i\}_{i=1,\ldots,n}$ is a linear coordinate system. If moreover s is an orthonormal linear coordinate system, then consider the constant symmetric 2-form

$$\omega_X := \sum_{i=1}^n \mathrm{d}s_i \otimes \mathrm{d}s_i \in \Omega^1_{\mathrm{const}}(X) \otimes_{\mathbb{R}} \Omega^1_{\mathrm{const}}(X) \subset \Omega^1(X) \otimes_{\mathcal{O}(X)} \Omega^1(X).$$

The form ω_X is independent of the orthonormal linear coordinate system. It encodes the global metric (the distance function dist_X) in the obvious way.

Suppose $f : X \to Y$ is a linear map between polyhedra. Then

$$f^*(\omega_Y) \in \Omega^1_{\mathrm{const}}(X) \otimes_{\mathbb{R}} \Omega^1_{\mathrm{const}}(X).$$

The map f is conformal if and only if $f^*(\omega_Y) = a\omega_X$ for some positive real number a, which we call the scaling factor. And f is a linear metric embedding if and only if $a = 1$.

Let X be a polyhedron. The linear functions on X, i.e. the elements of $\mathcal{O}_{\mathrm{lin}}(X)$, allow us to construct the convex hull of any subset $S \subset X$ (cf. Proposition 1.5.3). A sub-polyhedron Z of X is by definition the convex hull of a finite subset of X. Z inherits from X (by restriction) a structure of a polyhedron. The inclusion map $Z \to X$ is also called a linear metric embedding.

1.6   Piecewise Smooth Forms and Maps

Let X be an n-dimensional polyhedron (Definition 1.5.1). A p-dimensional linear simplex in X is by definition the image of an injective linear map $\tau : \Delta^p \to X$. A *linear triangulation of* X is a finite collection $\{X_j\}_{j \in J}$ of linear simplices in X, such that $X_j \neq X_k$ if $j \neq k$; each face of a simplex X_j is the simplex X_k for some $k \in J$; for any $j, k \in J$ the intersection $X_j \cap X_k$ is the simplex X_l for some $l \in J$; and $X = \bigcup_{j \in J} X_j$. We let $J_p := \{j \in J \mid \dim X_j = p\}$, so $J = \coprod_{0 \leq p \leq n} J_p$. The topological space

$$\mathrm{sk}_p \, T := \bigcup_{q \leq p} \bigcup_{j \in J_q} X_j$$

is called the *p-skeleton* of the triangulation T.

We shall have to use piecewise smooth differential forms on polyhedra. Suppose $T = \{X_j\}_{j \in J}$ is a linear triangulation of a polyhedron X. A *piecewise smooth differential p-form on X, relative to the triangulation T*, is a collection $\{\alpha_j\}_{j \in J}$ of differential forms, with $\alpha_j \in \Omega^p(X_j)$, such that for any inclusion $X_j \subset X_k$ of simplices one has $\alpha_j|_{X_k} = \alpha_k$. Let us denote by $\Omega^p_{\mathrm{pws}}(X; T)$ the set of piecewise smooth differential p-form on X relative to T, and

$$\Omega_{\mathrm{pws}}(X; T) := \bigoplus_{p=0}^{n} \Omega^p_{\mathrm{pws}}(X; T),$$

where $n := \dim X$. This is a DG algebra.

Next suppose $S = \{Y_k\}_{k \in K}$ is a linear triangulation of X which is a subdivision of T. This means that each simplex Y_k is contained in some simplex X_j. Take a piecewise smooth differential form $\alpha = \{\alpha_j\} \in \Omega^p_{\mathrm{pws}}(X; T)$. Then there is a unique piecewise smooth differential form $\beta = \{\beta_k\} \in \Omega^p_{\mathrm{pws}}(X; S)$ satisfying $\beta_k = \alpha_j|_{Y_k}$ for any inclusion $Y_k \subset X_j$. In this way we get a DG algebra homomorphism $\Omega_{\mathrm{pws}}(X; T) \to \Omega_{\mathrm{pws}}(X; S)$, which is actually injective. Since any two triangulations have a common subdivision, the DG algebras $\Omega_{\mathrm{pws}}(X; T)$ form a directed system.

Definition 1.6.1. Let X be a polyhedron. The *algebra of piecewise smooth differential forms on X* is the DG algebra

$$\Omega_{\mathrm{pws}}(X) := \lim_{T \to} \Omega_{\mathrm{pws}}(X; T),$$

where T runs over all linear triangulations of X.

This definition is very similar to Sullivan's PL forms; see [FHT].

Note that any element of $\Omega^p_{\mathrm{pws}}(X)$ is represented by some $\alpha \in \Omega^p_{\mathrm{pws}}(X; T)$, where T is a linear triangulation of X. Such a triangulation

T is called a *smoothing triangulation* for α. There are injective DG algebra homomorphisms

(1.6.2) $$\Omega(X) \to \Omega_{\mathrm{pws}}(X;T) \to \Omega_{\mathrm{pws}}(X).$$

It is an exercise to show that the cohomology of $\Omega(X)$ vanishes in positive degrees, and $\mathrm{H}^0 \, \Omega(X) = \mathbb{R}$ (recall that X is convex). Also the homomorphisms (1.6.2) are quasi-isomorphisms. (We shall not use these facts.)

Definition 1.6.3. Let X be a polyhedron, and let Y be a manifold (see Convention 1.4.7).

  (1) Let $T = \{X_j\}_{j \in J}$ be a linear triangulation of X. A map $f : X \to Y$ is called a *piecewise smooth map relative to T* if $f|_{X_j} : X_j \to Y$ is smooth for every $j \in J$.
  (2) A map $f : X \to Y$ is a *piecewise smooth map* if it is piecewise smooth relative to some linear triangulation T; and then we say that T is a *smoothing triangulation* for f.

Note that a piecewise smooth map $f : X \to \mathbf{A}^1$ is the same as an element of the $\mathbb{R}$-algebra $\mathcal{O}_{\mathrm{pws}}(X) := \Omega^0_{\mathrm{pws}}(X)$. Given a piecewise smooth map $f : X \to Y$, there is an induced DG algebra homomorphism

$$f^* : \Omega(Y) \to \Omega_{\mathrm{pws}}(X).$$

Next suppose X and Y are both polyhedra. Then we can talk about *piecewise linear maps* $f : X \to Y$, using linear triangulations of X as explained in the previous paragraph. Given a piecewise linear map $f : X \to Y$, there is an induced DG algebra homomorphism

$$f^* : \Omega_{\mathrm{pws}}(Y) \to \Omega_{\mathrm{pws}}(X).$$

As usual, if f is an embedding, then we write $\alpha|_X := f^*(\alpha)$ for $\alpha \in \Omega_{\mathrm{pws}}(Y)$.

Let $\alpha \in \Omega^p_{\mathrm{pws}}(X)$. The form α can be presented as follows. Choose a smoothing triangulation $T = \{X_j\}_{j \in J}$ for α; so that $\alpha|_{X_j} \in \Omega^p(X_j)$ for every j. Let $s = (s_1, \ldots, s_n)$ be a linear coordinate system on X. For any j let $\tilde{\alpha}_{j,i} \in \mathcal{O}(X_j)$ be the coefficients of $\alpha|_{X_j}$ with respect to s, as in Definition 1.3.4. Then

(1.6.4) $$\alpha|_{X_j} = \sum_i \tilde{\alpha}_{j,i} \cdot \mathrm{d}s_{i_1} \wedge \cdots \wedge \mathrm{d}s_{i_p} \in \Omega^p(X_j),$$

where the multi-index $i = (i_1, \ldots, i_p)$ runs through all strictly increasing elements in $\{1, \ldots, n\}^p$.

Let X be an n-dimensional polyhedron. Given a form $\alpha \in \Omega_{\text{pws}}(X)$ and a point $x \in X$, we say that x *is a smooth point of α* if there exists some n-dimensional simplex Y in X, such that $x \in \text{Int } Y$ and $\alpha|_Y \in \Omega(Y)$. The smooth locus of α, i.e. the set of all smooth points of α, is open and dense in X. Its complement, called the singular locus of α, is contained in a finite union of simplices of dimensions $< n$. Indeed, given any triangulation T that smooths α, the singular locus of α is contained in the $(n-1)$-skeleton of T.

Let $f : X \to Y$ be a piecewise smooth map from a polyhedron to a manifold. As above we can talk about the smooth locus of f, and whether $f|_Z : Z \to Y$ is smooth for some simplex $Z \subset X$.

1.7   Sobolev Norm

It shall be very convenient to have a bound for piecewise smooth forms on polyhedra and some of their derivatives.

Let X be a polyhedron, with orthonormal linear coordinate system $s = (s_1, \ldots, s_n)$. Given a smooth function $f \in \mathcal{O}(X)$ and a multi-index

$$i = (i_1, \ldots, i_q) \in \{1, \ldots, n\}^q,$$

we write

$$\partial_i f := \frac{\partial^q}{\partial s_{i_1} \cdots \partial s_{i_q}} f \in \mathcal{O}(X).$$

In the case $q = 0$, so the only sequence is the empty sequence $\varnothing$, we write $\partial_\varnothing f := f$. We refer to the operators ∂_i as the partial derivatives with respect to s.

Definition 1.7.1. Let X be a polyhedron of dimension n, and let s be an orthonormal linear coordinate system on X.

(1) Take a smooth differential form $\alpha \in \Omega^p(X)$. Consider the coefficients $\tilde{\alpha}_i$ of α with respect to s, as in Definition 1.3.4. Let ∂_j be the partial derivatives with respect to s. For a point $x \in X$ and a natural number q we define

$$\|\alpha\|_{\text{Sob};x,q} := \left(\sum_{i,j} (\partial_j \tilde{\alpha}_i)(x)^2 \right)^{1/2},$$

where the sum is on all $j \in \{1, \ldots, n\}^q$ and on all strictly increasing $i \in \{1, \ldots, n\}^p$.

(2) For $\alpha \in \Omega^p(X)$ we define

$$\|\alpha\|_{\text{Sob}} := \sup_{q=0,1,2} \; \sup_{x \in X} \|\alpha\|_{\text{Sob};x,q} .$$

(3) Take a piecewise smooth differential form $\alpha \in \Omega^p_{pws}(X)$. Let $\{X_j\}_{j \in J}$ be some smoothing triangulation for α, and let $\alpha_j := \alpha|_{X_j} \in \Omega^p(X_j)$. Define

$$\|\alpha\|_{Sob} := \sup_{j \in J} \|\alpha_j\|_{Sob} .$$

The number $\|-\|_{Sob}$ is called the *Sobolev norm to order* 2.

Remark 1.7.2. Even for a smooth function f, our Sobolev norm $\|f\|_{Sob}$ is not the same as the usual order 2 Sobolev norm from functional analysis, namely $\|f\|_s$ with $s = 2$. But it is in the "same spirit", and hence we use this name.

Proposition 1.7.3. *Let X be an n-dimensional polyhedron.*

(1) *For $\alpha \in \Omega^p(X)$, $x \in X$ and $q \in \mathbb{N}$, the number $\|\alpha\|_{Sob;x,q}$ is independent of the orthonormal linear coordinate system s on X. Hence the number $\|\alpha\|_{Sob}$ is also coordinate independent.*

(2) *Given $\alpha \in \Omega^p_{pws}(X)$, its Sobolev norm $\|\alpha\|_{Sob}$ is independent of the smoothing triangulation $\{X_j\}_{j \in J}$.*

(3) *Let Z be a sub-polyhedron of X and let $\alpha \in \Omega^p_{pws}(X)$. Then $\|\alpha|_Z\|_{Sob} \le \|\alpha\|_{Sob}$.*

Proof. (1) Let us denote by V the vector space of constant vector fields on X. This is a rank n vector space, endowed with a canonical inner product. The set $\{\frac{\partial}{\partial s_i}\}_{i=1,\dots,n}$ is an orthonormal basis for V.

For $p, q \in \mathbb{N}$ let us denote by $T^q(V)$ and $\bigwedge^p(V)$ the tensor power and exterior power of V, respectively. The vector space $T^q(V) \otimes \bigwedge^p(V)$ has an induced inner product.

For a strictly increasing sequence

$$i = (i_1, \dots, i_p) \in \{1, \dots, n\}^p$$

let us write

$$\pi_i := \frac{\partial}{\partial s_{i_1}} \wedge \cdots \wedge \frac{\partial}{\partial s_{i_p}} \in \bigwedge\nolimits^p(V).$$

Then the set $\{\partial_j \otimes \pi_i\}$, where the indices run over all $j \in \{1, \dots, n\}^q$ and all strictly increasing $i \in \{1, \dots, n\}^p$, is an orthonormal basis of $T^q(V) \otimes \bigwedge^p(V)$.

Now V is the linear dual of $\Omega^1_{const}(X)$, so π_i can be viewed as a linear map $\Omega^p(X) \to \mathcal{O}(X)$. Given $\alpha \in \Omega^p(X)$, its coefficients with respect to s satisfy $\tilde{\alpha}_i = \pi_i(\alpha)$. Thus

$$\partial_j \tilde{\alpha}_i = (\partial_j \circ \pi_i)(\alpha)$$

in $\mathcal{O}(X)$.

Suppose $s' = (s'_1, \ldots, s'_n)$ is another orthonormal linear coordinate system. Let us denote by ∂'_j, $\tilde{\alpha}'_i$ and π'_i the new operators and coefficient, namely those with respect to s'. The set $\{\partial'_j \otimes \pi'_i\}$ is also an orthonormal basis of $T^q(V) \otimes \bigwedge^p(V)$. So this set and the set $\{\partial_j \otimes \pi_i\}$ are related by a constant orthogonal matrix (of size $n^q \cdot \binom{n}{p}$). It follows that

$$\sum_{i,j} (\partial'_j \tilde{\alpha}'_i)(x)^2 = \sum_{i,j} (\partial_j \tilde{\alpha}_i)(x)^2$$

for any $x \in X$.

We see that $\|\alpha\|_{\text{Sob};x,q}$ is independent of coordinates. The assertion for $\|\alpha\|_{\text{Sob}}$ is now clear.

(2) The supremum on x can be restricted to a dense open set in each cell X_j of the triangulation $\{X_j\}_{j \in J}$. Therefore we get the same value by replacing the triangulation $\{X_j\}_{j \in J}$ with a refinement.

(3) Given Z, choose an orthonormal linear coordinate system on it, and extend it to an orthonormal linear coordinate system on X. Now it is clear. $\qquad\qquad\square$

1.8 Orientation and Integration

An orientation on a manifold means a choice of a volume form, up to multiplication by a positive smooth function. However in the case of a polyhedron we can normalize the volume form:

Definition 1.8.1. Let X be a polyhedron of dimension n. An *orientation* on X is a constant form $\text{or}(X) \in \Omega^n_{\text{const}}(X)$, such that

$$\text{or}(X) = ds_1 \wedge \cdots \wedge ds_n$$

for some orthonormal linear coordinate system $(s_1, \ldots, s_n)$ on X.

Note that if $(s'_1, \ldots, s'_n)$ is some other orthonormal linear coordinate system, then

$$\text{or}(X) = \pm ds'_1 \wedge \cdots \wedge ds'_n.$$

If the sign is $+$ then $(s'_1, \ldots, s'_n)$ is said to be *positively oriented*.

For the polyhedron $\mathbf{I}^n$ there is a standard orientation, coming from its embedding in $\mathbf{A}^n$. It is

$$\text{or}(\mathbf{I}^n) := dt_1 \wedge \cdots \wedge dt_n,$$

where $(t_1, \ldots, t_n)$ is the standard coordinate system on $\mathbf{A}^n$. Likewise for the polyhedron Δ^n: we let

$$\text{or}(\Delta^n) := dt_1 \wedge \cdots \wedge dt_n,$$

where $(t_0, \ldots, t_n)$ is the barycentric coordinate system on $\mathbf{A}^{n+1}$. For $n = 1, 2$ the orientations can be also described by arrows – see Figure 5.

Definition 1.8.2. Let X be an n-dimensional oriented polyhedron and let $\alpha \in \Omega^n(X)$. The *coefficient* of α is the function $\tilde{\alpha} \in \mathcal{O}(X)$ such that $\alpha = \tilde{\alpha} \cdot \operatorname{or}(X)$.

In other words, $\tilde{\alpha}$ is the coefficient of α with respect to any positively oriented orthonormal linear coordinate system, in the sense of Definition 1.3.4.

An orientation on X induces an orientation on the $(n-1)$-dimensional polyhedra that make up the boundary ∂X (by contracting the orientation volume form of X with a constant outer gradient to the face; cf. [Wa, Section 4.8]).

Let X be an n-dimensional oriented polyhedron. Suppose $\alpha = \{\alpha_j\}_{j \in J} \in \Omega^n_{\mathrm{pws}}(X; T)$, where $T = \{X_j\}_{j \in J}$ is some linear triangulation of X. Any n-dimensional simplex X_j inherits an orientation, and hence the integral $\int_{X_j} \alpha_j$ is well-defined. We let

$$\int_X \alpha := \sum_{j \in J_n} \int_{X_j} \alpha_j \in \mathbb{R} \, .$$

This integration is compatible with subdivisions, and thus we have a well-defined function

$$\int_X : \Omega^n_{\mathrm{pws}}(X) \to \mathbb{R}.$$

Theorem 1.8.3 (Stokes Theorem). *Let X be an oriented n-dimensional polyhedron, and let $\alpha \in \Omega^{n-1}_{\mathrm{pws}}(X)$. Then*

$$\int_X \mathrm{d}\alpha = \int_{\partial X} \alpha \, .$$

Proof. Choose a linear triangulation $T = \{X_j\}_{j \in J}$ of X that smooths α; so that $\alpha = \{\alpha_j\}_{j \in J}$ with $\alpha_j \in \Omega^{n-1}(X_j)$. Then by definition of $\int_{\partial X} \alpha$, and by cancellation due to opposite orientations of inner $(n-1)$-dimensional simplices, we get

$$\int_{\partial X} \alpha = \sum_{j \in J_n} \int_{\partial X_j} \alpha_j.$$

And the usual Stokes Theorem tells us that

$$\int_{\partial X_j} \alpha_j = \int_{X_j} \mathrm{d}\alpha_j \, .$$

$\square$

1.9   Piecewise Continuous Forms

For a polyhedron X, let us denote by $\mathcal{O}_{\mathrm{cont}}(X)$ the set of continuous $\mathbb{R}$-valued functions on X. This is a commutative $\mathbb{R}$-algebra containing $\mathcal{O}_{\mathrm{pws}}(X)$.

Definition 1.9.1. Let p be a natural number. By *piecewise continuous p-form* on X we mean an element $\gamma \in \Omega^p_{\mathrm{pwc}}(X)$, where we define

$$\Omega^p_{\mathrm{pwc}}(X) := \mathcal{O}_{\mathrm{cont}}(X) \otimes_{\mathcal{O}_{\mathrm{pws}}(X)} \Omega^p_{\mathrm{pws}}(X).$$

A piecewise continuous p-form γ can be represented as follows: there are continuous functions $f_1, \ldots, f_m \in \mathcal{O}_{\mathrm{cont}}(X)$, and piecewise smooth forms $\alpha_1, \ldots, \alpha_m \in \Omega^p_{\mathrm{pws}}(X)$, such that

$$(1.9.2) \qquad\qquad \gamma = \sum_{i=1}^{m} f_i \cdot \alpha_i.$$

For $p = 0$ we have $\Omega^p_{\mathrm{pwc}}(X) = \mathcal{O}_{\mathrm{cont}}(X)$ of course. But for $p > 0$ these forms are much more complicated, as we shall see below.

Warning: the exterior derivative $\mathrm{d}(\gamma)$ is not defined for $\gamma \in \Omega^p_{\mathrm{pwc}}(X)$; at least not as a piecewise continuous form (it is a distribution). The problem of course is that the functions f_i above cannot be derived.

However, given $\gamma \in \Omega^p_{\mathrm{pwc}}(X)$ and a piecewise linear map $g : Y \to X$ between polyhedra, the pullback

$$g^*(\gamma) \in \Omega^p_{\mathrm{pwc}}(Y)$$

is well defined. In terms of the expansion (1.9.2), the formula is

$$g^*(\gamma) = \sum_{i=1}^{m} g^*(f_i) \cdot g^*(\alpha_i).$$

Here $g^*(f_i) \in \mathcal{O}_{\mathrm{cont}}(Y)$ and $g^*(\alpha_i) \in \Omega^p_{\mathrm{pws}}(Y)$. If g is a closed embedding then we write $\gamma|_Y := g^*(\gamma)$ as usual.

The presentation (1.9.2) of $\gamma \in \Omega^p_{\mathrm{pwc}}(X)$ can be expanded further. Choose a linear triangulation $\{X_j\}_{j \in J}$ of X that smooths all the piecewise smooth forms $\alpha_1, \ldots, \alpha_m$. Also choose a linear coordinate system $s = (s_1, \ldots, s_n)$ on X. For any $i \in \{1, \ldots, m\}$ and $j \in J$ let $e_{i,j,k} \in \mathcal{O}(X_j)$ be the coefficients of the smooth form $\alpha_i|_{X_j}$ with respect to s, as in Definition 1.3.4. Then

$$(1.9.3) \qquad \gamma|_{X_j} = \sum_{i=1}^{m} \sum_{k} f_i \cdot e_{i,j,k} \cdot \mathrm{d}s_{k_1} \wedge \cdots \wedge \mathrm{d}s_{k_p} \in \Omega^p_{\mathrm{pwc}}(X_j),$$

where as usual the multi-index $k = (k_1, \ldots, k_p)$ runs through all strictly increasing elements in $\{1, \ldots, n\}^p$.

For top degree forms one can say more. Let $\gamma \in \Omega_{\mathrm{pwc}}^n(X)$. In this case the expansion (1.9.3) can be simplified to

$$(1.9.4) \qquad \gamma|_{X_j} = \sum_{i=1}^{m} f_i \cdot e_{i,j} \cdot \mathrm{d}s_1 \wedge \cdots \wedge \mathrm{d}s_n \in \Omega_{\mathrm{pwc}}^n(X_j),$$

with $e_{i,j} \in \mathcal{O}(X_j)$.

Now assume that the polyhedron X is oriented. We can choose the coordinate system $s = (s_1, \ldots, s_n)$ so that it is orthonormal and positively oriented; and then

$$\mathrm{or}(X) = \mathrm{d}s_1 \wedge \cdots \wedge \mathrm{d}s_n.$$

For any $j \in J_n$ we can integrate the continuous function $\sum_{i=1}^{m} f_i \cdot e_{i,j}$ on the oriented n-dimensional simplex X_j, obtaining

$$\int_{X_j} \gamma := \int_{X_j} \sum_{i=1}^{m} f_i \cdot e_{i,j} \cdot \mathrm{d}s_1 \wedge \cdots \wedge \mathrm{d}s_n \in \mathbb{R}.$$

Using this formula we define

$$(1.9.5) \qquad \int_X \gamma := \sum_{j \in J_n} \int_{X_j} \gamma \in \mathbb{R}.$$

Proposition 1.9.6. *For an n-dimensional oriented polyhedron X and a piecewise continuous form $\gamma \in \Omega_{\mathrm{pwc}}^n(X)$, the number $\int_X \gamma$ is independent of the presentation (1.9.4) of γ.*

We leave out the easy proof.

More generally, suppose X is an n-dimensional polyhedron (not necessarily oriented), Z is an oriented p-dimensional polyhedron, $\tau : Z \to X$ is a piecewise linear map, and $\gamma \in \Omega_{\mathrm{pwc}}^p(X)$. Then $\tau^*(\gamma) \in \Omega_{\mathrm{pwc}}^p(Z)$, and we can use (1.9.5) to integrate γ along τ as follows:

$$(1.9.7) \qquad \int_\tau \gamma := \int_Z \tau^*(\gamma) \in \mathbb{R}.$$

2 Estimates for the Nonabelian Exponential Map

2.1 The Exponential Map

We need some preliminary results on the exponential maps of Lie groups. Let G be a (finite dimensional) Lie group over $\mathbb{R}$, with Lie algebra $\mathfrak{g} = \mathrm{Lie}(G)$. The exponential map

$$\exp_G : \mathfrak{g} \to G$$

is an analytic map, satisfying $\exp_G(0) = 1$ and $\exp(-\alpha) = \exp(\alpha)^{-1}$. The exponential map is a diffeomorphism near $0 \in \mathfrak{g}$. Namely there is an open neighborhood $U_0(\mathfrak{g})$ of 0 in $\mathfrak{g}$, such that $V_0(G) := \exp_G(U_0(\mathfrak{g}))$ is open in G, and $\exp_G : U_0(\mathfrak{g}) \to V_0(G)$ is a diffeomorphism. Let $\log_G : V_0(G) \to U_0(\mathfrak{g})$ be the inverse of $\exp_G$. We call such $V_0(G)$ an *open neighborhood of 1 in G on which $\log_G$ is defined*. See [Va, Sections 2.10] for details.

The exponential map is functorial. Namely given a map $\phi : G \to H$ of Lie groups, the diagram

$$
\begin{array}{ccc}
\mathrm{Lie}(G) & \xrightarrow{\ \exp_G\ } & G \\
{\scriptstyle \mathrm{Lie}(\phi)}\big\downarrow & & \big\downarrow{\scriptstyle \phi} \\
\mathrm{Lie}(H) & \xrightarrow{\ \exp_H\ } & H
\end{array}
$$

is commutative.

When there is no danger of confusion we write $\exp$ and $\log$ instead of $\exp_G$ and $\log_G$, respectively.

The product in G is denoted by $\cdot$, and the Lie bracket in $\mathfrak{g}$ is denoted by $[-,-]$. Given a finite sequence $(g_1, \ldots, g_m)$ of elements in the group G, we write

$$(2.1.1) \qquad \prod_{i=1}^{m} g_i := g_1 \cdot g_2 \cdots g_m \in G .$$

It is a basic fact that if $\alpha_1, \ldots, \alpha_m \in \mathfrak{g}$ are *commuting* elements, then

$$\prod_{i=1}^{m} \exp(\alpha_i) = \exp\left(\sum_{i=1}^{m} \alpha_i\right).$$

The next theorem lists several estimates for the discrepancy when the elements do not necessarily commute.

Theorem 2.1.2. *Let G be a Lie group, with Lie algebra $\mathfrak{g}$. Let $V_0(G)$ be an open neighborhood of 1 in G on which $\log$ is well-defined, and let $\|-\|$ be a euclidean norm on $\mathfrak{g}$. Then there are real constants $\epsilon_0(G)$ and $c_0(G)$ with the following properties:*

(i) $0 < \epsilon_0(G) \le 1$ *and* $0 \le c_0(G)$.

(ii) *Let* $\alpha_1, \ldots, \alpha_m \in \mathfrak{g}$ *be such that* $\sum_{i=1}^m \|\alpha_i\| < \epsilon_0(G)$. *Then*

$$\prod_{i=1}^m \exp(\alpha_i) \in V_0(G),$$

$$\left\| \log\left(\prod_{i=1}^m \exp(\alpha_i)\right) \right\| \le c_0(G) \cdot \sum_{i=1}^m \|\alpha_i\|$$

and

$$\left\| \log\left(\prod_{i=1}^m \exp(\alpha_i)\right) - \sum_{i=1}^m \alpha_i \right\| \le c_0(G) \cdot \left(\sum_{i=1}^m \|\alpha_i\|\right)^2.$$

(iii) *Let* $\alpha_1, \alpha_2 \in \mathfrak{g}$ *be such that* $\|\alpha_1\| + \|\alpha_2\| < \epsilon_0(G)$. *Then*

$$\|[\alpha_1, \alpha_2]\| \le c_0(G) \cdot \|\alpha_1\| \cdot \|\alpha_2\|,$$

$$\left\| \log\big(\exp(\alpha_1) \cdot \exp(\alpha_2)\big) - \left(\alpha_1 + \alpha_2 + \tfrac{1}{2}[\alpha_1, \alpha_2]\right) \right\|$$
$$\le c_0(G) \cdot \|\alpha_1\| \cdot \|\alpha_2\| \cdot \left(\|\alpha_1\| + \|\alpha_2\|\right)$$

and

$$\left\| \log\big(\exp(\alpha_1) \cdot \exp(\alpha_2) \cdot \exp(\alpha_1)^{-1} \cdot \exp(\alpha_2)^{-1}\big) - [\alpha_1, \alpha_2] \right\|$$
$$\le c_0(G) \cdot \|\alpha_1\| \cdot \|\alpha_2\| \cdot \left(\|\alpha_1\| + \|\alpha_2\|\right).$$

(iv) *Let* $\alpha_1, \ldots, \alpha_m, \beta_1, \ldots, \beta_m \in \mathfrak{g}$ *be such that*

$$\sum_{i=1}^m \left(\|\alpha_i\| + \|\beta_i\|\right) < \epsilon_0(G).$$

Then

$$\left\| \log\left(\prod_{i=1}^m \exp(\alpha_i + \beta_i)\right) - \log\left(\prod_{i=1}^m \exp(\alpha_i)\right) - \sum_{i=1}^m \beta_i \right\|$$
$$\le c_0(G) \cdot \left(\sum_{i=1}^m (\|\alpha_i\| + \|\beta_i\|)\right) \cdot \left(\sum_{i=1}^m \|\beta_i\|\right)$$

and

$$\left\| \log\left(\prod_{i=1}^m \exp(\alpha_i + \beta_i)\right) - \log\left(\prod_{i=1}^m \exp(\alpha_i)\right) \right\|$$
$$\le c_0(G) \cdot \left(\sum_{i=1}^m \|\beta_i\|\right).$$

(v) *Let $\alpha_1,\ldots,\beta_m$ be as in property* (iv), *and assume moreover that* $[\alpha_i,\alpha_j]$
$= 0$ *for all* i,j. *Then*

$$\| \log(\textstyle\prod_{i=1}^m \exp(\alpha_i + \beta_i)) - \sum_{i=1}^m (\alpha_i + \beta_i) \|$$
$$\leq c_0(G) \cdot (\textstyle\sum_{i=1}^m (\|\alpha_i\| + \|\beta_i\|)) \cdot (\sum_{i=1}^m \|\beta_i\|).$$

The theorem is proved at the end of Section 2.5, after some preparations.

The constants $c_0(G)$ and $\epsilon_0(G)$ are called a *commutativity constant* and a *convergence radius* for G respectively. (If G is abelian one can take $c_0(G) = 0$.)

2.2 The CBH Theorem

There is an element $F(x,y)$ in the completed free Lie algebra over $\mathbb{Q}$ in the variables x,y, called the *Hausdorff series*. See [Bo, Sections II.6 and II.7], where the letter H is used instead of F. For any $i,j \geq 0$ let us denote by $F_{i,j}(x,y)$ the homogeneous component of $F(x,y)$ of degree i in x and degree j in y. So

$$F(x,y) = \sum_{i,j\geq 0} F_{i,j}(x,y).$$

Now consider a Lie group G as before, with Lie algebra $\mathfrak{g}$. Choose a euclidean norm $\|-\|$ on the vector space $\mathfrak{g}$. Given elements $\alpha,\beta \in \mathfrak{g}$ we can evaluate $F_{i,j}(x,y)$ on them, obtaining an element $F_{i,j}(\alpha,\beta) \in \mathfrak{g}$.

The *Campbell-Baker-Hausdorff Theorem* says that there is an open neighborhood U_1 of 0 in $\mathfrak{g}$ (its size depending on the norm) such that the series $F(\alpha,\beta)$ converges absolutely and uniformly for $\alpha,\beta \in U_1$, and moreover

(2.2.1) $$\exp(F(\alpha,\beta)) = \exp(\alpha) \cdot \exp(\beta).$$

This assertion is not explicit in the book [Bo]; it requires combining various scattered results in Sections II.7 and III.4 of [Bo]. An explicit statement of the CBH Theorem is [Va, Theorem 2.15.4]. However the treatment of the structure of the Hausdorff series is not sufficiently detailed in [Va].

For a positive number r we denote by $U(r)$ the open ball of radius r and center 0 in $\mathfrak{g}$, and by $\overline{U}(r)$ its closure (the closed ball).

Lemma 2.2.2. *There are constants ϵ_1 and c_1, such that $0 < \epsilon_1$, $U(\epsilon_1) \subset U_1$, $0 \leq c_1$, and for every $\alpha,\beta \in U(\epsilon_1)$ the inequality*

$$\| F(\alpha,\beta) - (\alpha + \beta + \tfrac{1}{2}[x,y]) \| \leq c_1 \cdot \|\alpha\| \cdot \|\beta\| \cdot (\|\alpha\| + \|\beta\|)$$

holds.

Proof. Let us denote by $f_{i,j}$ the norm of the bi-homogeneous function

$$F_{i,j} : \mathfrak{g} \times \mathfrak{g} \to \mathfrak{g}.$$

Namely

$$f_{i,j} := \sup \{\|F_{i,j}(\alpha, \beta)\| \mid \|\alpha\|, \|\beta\| \leq 1\}.$$

So for any $\alpha, \beta \in \mathfrak{g}$ one has

$$\|F_{i,j}(\alpha, \beta)\| \leq f_{i,j} \cdot \|\alpha\|^i \cdot \|\beta\|^j.$$

In [Bo, Section II.7.2] it is shown that there is a positive number ϵ_1, such that the series $\sum_{i,j \geq 0} f_{i,j} \epsilon_1^{i+j}$ converges (to a finite sum). By shrinking ϵ_1 if necessary, we may assume that $\overline{U}(\epsilon_1) \subset U_1$.

Now in [Bo, Section II.6.4] it is proved that

$$F_{1,0}(x,y) = x, \ F_{0,1}(x,y) = y, \ F_{1,1}(x,y) = \tfrac{1}{2}[x,y]$$

and

$$F_{i,0}(x,y) = F_{0,j}(x,y) = 0 \ \text{ for } \ i, j \neq 1.$$

Thus for $\alpha, \beta \in U(\epsilon_1)$ we have

$$F(\alpha, \beta) - (\alpha + \beta + \tfrac{1}{2}[x,y]) = \sum_{(i,j) \in I} F_{i,j}(\alpha, \beta),$$

where

$$I := \{(i,j) \mid i, j \geq 1 \text{ and } i + j \geq 3\}.$$

Now if $(i,j) \in I$ and $0 \leq a, b \leq \epsilon_1$, then

$$a^i b^j \leq ab(a+b) \cdot \epsilon_1^{i+j-3}.$$

Therefore for $\alpha, \beta \in U(\epsilon_1)$, with $a := \|\alpha\|$ and $b := \|\beta\|$, we have the estimate

$$\left\| \sum_{(i,j) \in I} F_{i,j}(\alpha, \beta) \right\| \leq \sum_{(i,j) \in I} \| F_{i,j}(\alpha, \beta) \|$$

$$\leq \sum_{(i,j) \in I} f_{i,j} \cdot a^i \cdot b^j \leq ab(a+b) \cdot \epsilon_1^{-3} \cdot \sum_{(i,j) \in I} f_{i,j} \cdot \epsilon_1^{i+j}.$$

Taking the number

$$c_1 := \epsilon_1^{-3} \cdot \sum_{(i,j) \in I} f_{i,j} \cdot \epsilon_1^{i+j}$$

does the trick. $\qquad\qquad\qquad\qquad\qquad\qquad\qquad\qquad\qquad\qquad\square$

2.3 Calculations with a Few Elements

We continue with the chosen open set $V_0(G) \subset G$ and norm $\|-\|$ on the Lie algebra $\mathfrak{g}$. In this section we translate the estimate for the CBH series to estimates on exp and log. Recall our notation: $U(r)$ is the open ball of positive radius r in $\mathfrak{g}$, and $\overline{U}(r)$ is the closed ball.

Lemma 2.3.1. *There exist constants* $\epsilon_2, c_2 \in \mathbb{R}$ *such that*

$$0 < \epsilon_2 \leq 1 \text{ and } 1 \leq c_2,$$

and such the following formulas hold for every $\alpha, \beta \in U(\epsilon_2)$.

(2.3.2) $$\exp(\alpha) \cdot \exp(\beta) \in V_0(G),$$

(2.3.3) $$\| [\alpha, \beta] \| \leq c_2 \cdot \|\alpha\| \cdot \|\beta\|,$$

and

(2.3.4)
$$\| \log(\exp(\alpha) \cdot \exp(\beta)) - (\alpha + \beta + \tfrac{1}{2}[\alpha, \beta]) \|$$
$$\leq c_2 \cdot \|\alpha\| \cdot \|\beta\| \cdot (\|\alpha\| + \|\beta\|).$$

Proof. The existence of a positive ϵ_2 making formula (2.3.2) true is easy (due to continuity). We can assume $\epsilon_2 \leq \min(1, \epsilon_1)$.

The existence of a number c_2 making inequality (2.3.3) valid is also easy, because the function $\| [\alpha, \beta] \|$ is bilinear and $\mathfrak{g}$ is finite dimensional. We can assume that $c_2 \geq \max(1, c_1)$.

Finally, the validity of formula (2.3.4) comes from Lemma 2.2.2 and the CBH formula (2.2.1). □

Lemma 2.3.5. *There is a real number* c_3 *such that* $c_3 \geq c_2$, *and for any* $\alpha_1, \alpha_2, \beta_1, \beta_2 \in \mathfrak{g}$ *satisfying* $\|\alpha_i\| + \|\beta_i\| < \epsilon_2$ *the inequality*

(2.3.6)
$$\| \log(\exp(\alpha_1 + \beta_1) \cdot \exp(\alpha_2 + \beta_2)) - \log(\exp(\alpha_1) \cdot \exp(\alpha_2)) \|$$
$$\leq c_3 \cdot (\|\beta_1\| + \|\beta_2\|)$$

holds.

Proof. Given $\alpha \in \overline{U}(\epsilon_2)$ we have a smooth (in fact analytic) function $f_\alpha : U(\epsilon_2) \to \mathfrak{g}$ defined by

$$f_\alpha(\beta) := \log(\exp(\alpha) \cdot \exp(\beta)).$$

Let $\mathfrak{g}^*$ be the dual space of $\mathfrak{g}$, and let $d_\beta f_\alpha \in \mathfrak{g}^*$ be the derivative of f_α at β. Then Taylor expansion of f_α around β gives us

(2.3.7) $$f_\alpha(\beta + \gamma) = f_\alpha(\beta) + (d_\beta f_\alpha)(\gamma) + (R_\beta^{\geq 2} f_\alpha)(\gamma)$$

for any $\gamma \in \overline{U}(\epsilon_2)$. Here $R^{\geq 2}_{\beta} f_{\alpha}$ is just the remainder, in other words the higher order part of the Taylor expansion. The linear term $(d_{\beta} f_{\alpha})(\gamma)$ can be estimated by

$$\| (d_{\beta} f_{\alpha})(\gamma) \| \leq a_1 \cdot \|\gamma\|$$

for a suitable positive number a_1. The remainder $(R^{\geq 2}_{\beta} f_{\alpha})(\gamma)$ can be estimated by

$$\| (R^{\geq 2}_{\beta} f_{\alpha})(\gamma) \| \leq a_2 \cdot \|\gamma\|^2$$

for some positive number a_2.

These inequalities tell us that if $\alpha, \beta, \gamma \in U(\epsilon_2)$ then

(2.3.8)
$$\begin{aligned}
\| \log(\exp(\alpha) \cdot \exp(\beta + \gamma)) &- \log(\exp(\alpha) \cdot \exp(\beta)) \| \\
= \| \log(f_{\alpha}(\beta + \gamma)) &- \log(f_{\alpha}(\beta)) \| \\
\leq a_1 \cdot \|\gamma\| + a_2 \cdot \|\gamma\|^2 &\leq (a_1 + a_2) \cdot \|\gamma\|.
\end{aligned}$$

In a similar way we look at the function $g_{\beta}(\alpha) := f_{\alpha}(\beta)$, and we obtain constants b_1, b_2 such that

(2.3.9)
$$\| \log(\exp(\alpha + \gamma) \cdot \exp(\beta)) - \log(\exp(\alpha) \cdot \exp(\beta)) \| \leq (b_1 + b_2) \cdot \|\gamma\|.$$

Therefore by taking

$$c_3 := \max\{c_2, a_1 + a_1, b_1 + b_2\}$$

all conditions are satisfied. $\qquad \square$

Lemma 2.3.10. *There exists a constant $c_4 \geq c_3$ such that the following inequalities hold for any $\alpha, \beta \in \mathfrak{g}$ satisfying $\|\alpha\|, \|\beta\| < \epsilon_2$.*

$$\| \log(\exp(\alpha) \cdot \exp(\beta)) - (\alpha + \beta) \| \leq c_4 \cdot \|\alpha\| \cdot \|\beta\|$$

and

$$\| \log(\exp(\alpha) \cdot \exp(\beta)) \| \leq \|\alpha\| + \|\beta\| + c_4 \cdot \|\alpha\| \cdot \|\beta\|.$$

Proof. From Lemma 2.3.1 and the triangle inequality we get

$$\begin{aligned}
\| \log(\exp(\alpha) \cdot \exp(\beta)) &- (\alpha + \beta) \| \\
\leq c_2 \cdot \|\alpha\| \cdot \|\beta\| \cdot (\|\alpha\| + \|\beta\|) &+ \tfrac{1}{2} c_2 \cdot \|\alpha\| \cdot \|\beta\| \\
\leq 3 c_2 \cdot \|\alpha\| \cdot \|\beta\|.
\end{aligned}$$

This proves the first inequality, with $c_4 := \max(c_3, 3c_2)$. The second inequality is an immediate consequence. $\qquad \square$

2.4 Calculations with Many Elements

Let us define

(2.4.1) $$\epsilon_4 := \tfrac{1}{12} c_4^{-1} \cdot \epsilon_2.$$

Lemma 2.4.2. *Let $\alpha_1, \ldots, \alpha_m \in \mathfrak{g}$ be such that $\sum_{i=1}^m \|\alpha_i\| < \epsilon_4$. Then the relation and inequalities below hold:*

(2.4.3) $$\prod_{i=1}^m \exp(\alpha_i) \in V_0(G),$$

(2.4.4) $$\big\| \log\big(\textstyle\prod_{i=1}^m \exp(\alpha_i)\big) \big\| \le \tfrac{3}{2} \sum_{i=1}^m \|\alpha_i\|$$

and

(2.4.5) $$\big\| \log\big(\textstyle\prod_{i=1}^m \exp(\alpha_i)\big) - \sum_{i=1}^m \alpha_i \big\| \le c_4 \cdot \big(\sum_{i=1}^m \|\alpha_i\|\big)^2.$$

Proof. Recall that $\epsilon_2 \le 1$ and $c_4 \ge 1$.

For $m = 1$ all is clear. Take $m = 2$. Then since $\|\alpha_i\| < \epsilon_2$ we have

$$\exp(\alpha_1) \cdot \exp(\alpha_2) \in V_0(G)$$

by Lemma 2.3.1. Next, since $\|\alpha_1\| \le \tfrac{1}{6} c_4^{-1}$, it follows that

$$c_4 \cdot \|\alpha_1\| \cdot \|\alpha_2\| \le \tfrac{1}{6} \|\alpha_2\|.$$

From this and Lemma 2.3.10 we see that

$$\big\| \exp(\alpha_1) \cdot \exp(\alpha_2) \big\| \le \|\alpha_1\| + \|\alpha_2\| + c_4 \cdot \|\alpha_1\| \cdot \|\alpha_2\| \le \tfrac{3}{2} \cdot (\|\alpha_1\| + \|\alpha_2\|).$$

Hence (2.4.4) holds for $m = 2$. Again using Lemma 2.3.10 we have

$$\big\| \log\big(\exp(\alpha_1) \cdot \exp(\alpha_2)\big) - (\alpha_1 + \alpha_2) \big\|$$
$$\le c_4 \cdot \|\alpha_1\| \cdot \|\alpha_2\| \le c_4 \cdot (\|\alpha_1\| + \|\alpha_2\|)^2.$$

This finishes the case $m = 2$.

Now assume the assertions are true for $m \ge 2$, and consider $\alpha_1, \ldots,$ $\alpha_{m+1} \in \mathfrak{g}$ such that $\sum_{i=1}^{m+1} \|\alpha_i\| < \epsilon_4$. Define

$$\beta := \log\big(\textstyle\prod_{i=1}^m \exp(\alpha_i)\big),$$

so that

$$\prod_{i=1}^{m+1} \exp(\alpha_i) = \exp(\beta) \cdot \exp(\alpha_{m+1}).$$

By the induction hypothesis, i.e. inequality (2.4.4) for m, we have

(2.4.6) $$\|\beta\| \le \tfrac{3}{2} \sum_{i=1}^m \|\alpha_i\| \le \tfrac{3}{2} \sum_{i=1}^{m+1} \|\alpha_i\| \le \tfrac{3}{2} \cdot \tfrac{1}{6} c_4^{-1} \cdot \epsilon_2 = \tfrac{1}{4} c_4^{-1} \cdot \epsilon_2.$$

This implies $\|\beta\| < \epsilon_2$. Since we also have $\|\alpha_{m+1}\| < \epsilon_2$, Lemma 2.3.1 says that

$$\exp(\beta) \cdot \exp(\alpha_{m+1}) \in V_0(G).$$

This verifies (2.4.3) for $m + 1$.

Using Lemma 2.3.1 again we see that

$$\| \log(\exp(\beta) \cdot \exp(\alpha_{m+1})) \| \leq \|\beta\| + \|\alpha_{m+1}\| + c_4 \cdot \|\beta\| \cdot \|\alpha_{m+1}\|.$$

Because we also have the inequality $c_4 \cdot \|\beta\| \leq \frac{1}{2}$, we conclude that

$$\|\beta\| + \|\alpha_{m+1}\| + c_4 \cdot \|\beta\| \cdot \|\alpha_{m+1}\| \leq \tfrac{3}{2} \cdot \|\beta\| + \tfrac{3}{2} \cdot \|\alpha_{m+1}\| \leq \tfrac{3}{2} \cdot \sum_{i=1}^{m+1} \|\alpha_i\|.$$

Thus inequality (2.4.4) holds for $m + 1$.

According to Lemma 2.3.10 we have

$$\| \log(\exp(\beta) \cdot \exp(\alpha_{m+1})) - (\beta + \alpha_{m+1}) \| \leq c_4 \cdot \|\beta\| \cdot \|\alpha_{m+1}\|.$$

The induction assumption for inequality (2.4.5) says that

$$\| \beta - \sum_{i=1}^m \alpha_i \| \leq c_4 \cdot \left(\sum_{i=1}^m \|\alpha_i\|\right)^2.$$

Combining these with the inequality (2.4.6) we obtain

$$\| \log(\exp(\beta) \cdot \exp(\alpha_{m+1})) - \sum_{i=1}^{m+1} \alpha_i \|$$
$$\leq \| \log(\exp(\beta) \cdot \exp(\alpha_{m+1})) - (\beta + \alpha_{m+1}) \|$$
$$+ \| (\beta + \alpha_{m+1}) - \sum_{i=1}^{m+1} \alpha_i \|$$
$$\leq c_4 \cdot \|\beta\| \cdot \|\alpha_{m+1}\| + c_4 \cdot \left(\sum_{i=1}^m \|\alpha_i\|\right)^2$$
$$\leq c_4 \cdot \left(\left(\tfrac{3}{2} \sum_{i=1}^m \|\alpha_i\|\right) \cdot \|\alpha_{m+1}\| + \left(\sum_{i=1}^m \|\alpha_i\|\right)^2 \right)$$
$$\leq c_4 \cdot \left(\sum_{i=1}^{m+1} \|\alpha_i\|\right)^2.$$

This proves (2.4.5). □

Lemma 2.4.7. *Let* $\alpha_1, \ldots, \alpha_m, \beta_1, \ldots, \beta_m \in \mathfrak{g}$ *be such that*

$$\sum_{i=1}^m (\|\alpha_i\| + \|\beta_i\|) < \epsilon_4.$$

Then

$$\| \log\left(\prod_{i=1}^m (\exp(\alpha_i) \cdot \exp(\beta_i))\right)$$
$$- \log\left((\prod_{i=1}^m \exp(\alpha_i)) \cdot (\prod_{i=1}^m \exp(\beta_i))\right) \|$$
$$\leq 4c_4^2 \cdot \left(\sum_{i=1}^m \|\alpha_i\|\right) \cdot \left(\sum_{i=1}^m \|\beta_i\|\right)$$

and

$$\left\| \log\left(\prod_{i=1}^{m} \exp(\alpha_i + \beta_i)\right) - \log\left(\prod_{i=1}^{m} (\exp(\alpha_i) \cdot \exp(\beta_i))\right) \right\|$$
$$\leq 2c_4^2 \cdot \sum_{i=1}^{m} \|\alpha_i\| \cdot \|\beta_i\|.$$

Proof. For the first inequality we note that for any i, j one has

$$\left\| \log\left(\exp(\beta_j) \cdot \exp(\alpha_i)\right) - \log\left(\exp(\alpha_i) \cdot \exp(\beta_j)\right) \right\|$$
$$\leq 2c_2 \cdot \|\alpha_i\| \cdot \|\beta_j\| \cdot (\|\alpha_i\| + \|\beta_j\|) + c_2 \cdot \|\alpha_i\| \cdot \|\beta_j\|$$
$$\leq 2c_2 \cdot \|\alpha_i\| \cdot \|\beta_j\|.$$

This is due to Lemma 2.3.1. Now for our inequality we have to move all the $\exp(\beta_j)$ in the product across all the $\exp(\alpha_i)$. According Lemma 2.3.5 the "cost" of each such move is at most $2c_3 \cdot 2c_2 \cdot \|\alpha_i\| \cdot \|\beta_j\|$. So the total "cost" is at most

$$\sum_{i,j=1}^{m} 2c_3 \cdot 2c_2 \cdot \|\alpha_i\| \cdot \|\beta_j\| = 4c_2c_3 \cdot \left(\sum_{i=1}^{m} \|\alpha_i\|\right) \cdot \left(\sum_{j=1}^{m} \|\beta_j\|\right).$$

Since $c_4 \geq c_2, c_3$ this proves the first inequality.

For the second inequality we note that, due to Lemma 2.3.10, we have

$$\left\| \log\left(\exp(\alpha_i + \beta_i)\right) - \log\left(\exp(\alpha_i) \cdot \exp(\beta_i)\right) \right\| \leq c_4 \cdot \|\alpha_i\| \cdot \|\beta_i\|.$$

Since we have to make such a change for every i, according to Lemma 2.3.5 the total "cost" is at most $2c_3c_4 \cdot \sum_{i=1}^{m} \|\alpha_i\| \cdot \|\beta_i\|$. □

Lemma 2.4.8. *There exists a constant c_5 such that $c_5 \geq \max(c_4, \frac{3}{2})$, and for every $\alpha_1, \ldots, \alpha_m, \beta_1, \ldots, \beta_m \in \mathfrak{g}$ satisfying*

$$\sum_{i=1}^{m} \left(\|\alpha_i\| + \|\beta_i\|\right) < \epsilon_4$$

one has:

(2.4.9)
$$\left\| \log\left(\prod_{i=1}^{m} \exp(\alpha_i + \beta_i)\right) - \log\left(\prod_{i=1}^{m} \exp(\alpha_i)\right) - \sum_{i=1}^{m} \beta_i \right\|$$
$$\leq c_5 \cdot \left(\sum_{i=1}^{m} (\|\alpha_i\| + \|\beta_i\|)\right) \cdot \left(\sum_{i=1}^{m} \|\beta_i\|\right)$$

and

(2.4.10)
$$\left\| \log\left(\prod_{i=1}^{m} \exp(\alpha_i + \beta_i)\right) - \log\left(\prod_{i=1}^{m} \exp(\alpha_i)\right) \right\|$$
$$\leq c_5 \cdot \left(\sum_{i=1}^{m} \|\beta_i\|\right).$$

If moreover $[\alpha_i, \alpha_j] = 0$ for all i, j, then

(2.4.11)
$$\left\| \log\left(\prod_{i=1}^{m} \exp(\alpha_i + \beta_i)\right) - \sum_{i=1}^{m} (\alpha_i + \beta_i) \right\|$$
$$\leq c_5 \cdot \left(\sum_{i=1}^{m} (\|\alpha_i\| + \|\beta_i\|)\right) \cdot \left(\sum_{i=1}^{m} \|\beta_i\|\right).$$

Proof. According to Lemma 2.4.7 we have
(2.4.12)
$$\| \log(\prod_{i=1}^{m} \exp(\alpha_i + \beta_i)) - \log\Big((\prod_{i=1}^{m} \exp(\alpha_i)) \cdot (\prod_{i=1}^{m} \exp(\beta_i)) \Big) \|$$
$$\leq 2c_4^2 \cdot \sum_{i=1}^{m} \|\alpha_i\| \cdot \|\beta_i\| .$$

By Lemma 2.4.2 we know that

(2.4.13) $$\| \log(\prod_{i=1}^{m} \exp(\beta_i)) - \sum_{i=1}^{m} \beta_i \| \leq c_4 \cdot \left(\sum_{i=1}^{m} \|\beta_i\| \right)^2 .$$

Combining Lemmas 2.3.10 and 2.4.2 we get

(2.4.14)
$$\| \log\Big((\prod_{i=1}^{m} \exp(\alpha_i)) \cdot (\prod_{i=1}^{m} \exp(\beta_i)) \Big)$$
$$- \Big(\log(\prod_{i=1}^{m} \exp(\alpha_i)) + \log(\prod_{i=1}^{m} \exp(\beta_i)) \Big) \|$$
$$\leq c_4 \cdot \| \log(\prod_{i=1}^{m} \exp(\alpha_i)) \| \cdot \| \log(\prod_{i=1}^{m} \exp(\beta_i)) \|$$
$$\leq c_4 \cdot (\tfrac{3}{2} \sum_{i=1}^{m} \|\alpha_i\|) \cdot (\tfrac{3}{2} \sum_{i=1}^{m} \|\beta_i\|)$$
$$= \tfrac{9}{4}c_4 \cdot (\sum_{i=1}^{m} \|\alpha_i\|) \cdot (\sum_{i=1}^{m} \|\beta_i\|) .$$

Putting all these together we obtain the inequality

(2.4.15)
$$\| \log(\prod_{i=1}^{m} \exp(\alpha_i + \beta_i)) - \log(\prod_{i=1}^{m} \exp(\alpha_i)) - \sum_{i=1}^{m} \beta_i \|$$
$$\leq 2c_4^2 \cdot (\sum_{i=1}^{m} \|\alpha_i\|) \cdot (\sum_{i=1}^{m} \|\beta_i\|)$$
$$+ \tfrac{9}{4}c_4^2 \cdot (\sum_{i=1}^{m} \|\alpha_i\|) \cdot (\sum_{i=1}^{m} \|\beta_i\|)$$
$$+ c_4 \cdot (\sum_{i=1}^{m} \|\beta_i\|)^2$$
$$\leq 6c_4^2 \cdot (\sum_{i=1}^{m} (\|\alpha_i\| + \|\beta_i\|)) \cdot (\sum_{i=1}^{m} \|\beta_i\|) .$$

This proves inequality (2.4.9) with $c_5 := 7c_4^2$.

Inequality (2.4.10) follows easily from (2.4.15), once we notice that

$$6c_4^2 \cdot (\sum_{i=1}^{m} (\|\alpha_i\| + \|\beta_i\|)) \leq 6c_4^2 \cdot 2\epsilon_3 \leq c_4 .$$

When $[\alpha_i, \alpha_j] = 0$ for all i, j we have

$$\log(\prod_{i=1}^{m} \exp(\alpha_i)) = \sum_{i=1}^{m} \alpha_i.$$

So inequality (2.4.10) holds. $\square$

2.5 Final Touches

Lemma 2.5.1. *There is a real number c_6 such that $c_6 \geq c_5$, and for any $\alpha, \beta \in \mathfrak{g}$ with $\|\alpha\|, \|\beta\| < \epsilon_4$ the inequality*

$$\| \log(\exp(\alpha) \cdot \exp(\beta) \cdot \exp(\alpha)^{-1} \cdot \exp(\beta)^{-1}) - [\alpha, \beta] \|$$
$$\leq c_6 \cdot \|\alpha\| \cdot \|\beta\| \cdot (\|\alpha\| + \|\beta\|)$$

(2.5.2)

holds.

Proof. Let us write
$$a := \exp(\alpha), \quad b := \exp(\beta),$$
$$\gamma^+ := \alpha + \beta + \tfrac{1}{2}[\alpha, \beta], \quad \gamma^- := -\alpha - \beta + \tfrac{1}{2}[\alpha, \beta],$$
$$\delta^+ := \log(a \cdot b) - \gamma^+, \quad \delta^- := \log(a^{-1} \cdot b^{-1}) - \gamma^-.$$

According to Lemma 2.3.1 we have
$$\|\delta^+\|, \|\delta^-\| \leq c_2 \cdot \|\alpha\| \cdot \|\beta\| \cdot (\|\alpha\| + \|\beta\|)$$

and
$$\|\gamma^+\|, \|\gamma^-\| \leq \|\alpha\| + \|\beta\| + \tfrac{1}{2}c_2 \cdot \|\alpha\| \cdot \|\beta\| \leq (1 + c_2)(\|\alpha\| + \|\beta\|).$$

Now
$$a \cdot b = \exp(\gamma^+ + \delta^+)$$
and
$$a^{-1} \cdot b^{-1} = \exp(\gamma^- + \delta^-).$$
So by Lemma 2.3.5 we have
$$\| \log((a \cdot b) \cdot (a^{-1} \cdot b^{-1})) - \log(\exp(\gamma^+) \cdot \exp(\gamma^-)) \|$$
$$\leq c_3 \cdot (\|\delta^+\| + \|\delta^-\|)$$
$$\leq 2c_2c_3 \cdot \|\alpha\| \cdot \|\beta\| \cdot (\|\alpha\| + \|\beta\|).$$

On the other hand, since $\alpha + \beta$ and $-\alpha - \beta$ commute, by formula (2.4.11) in Lemma 2.4.8 we have
$$\| \log(\exp(\gamma^+) \cdot \exp(\gamma^-)) - [\alpha, \beta] \|$$
$$= \| \log(\exp((\alpha + \beta) + \tfrac{1}{2}[\alpha, \beta]) \cdot \exp((-\alpha - \beta) + \tfrac{1}{2}[\alpha, \beta]))$$
$$- ((\alpha + \beta) + \tfrac{1}{2}[\alpha, \beta] + (-\alpha - \beta) + \tfrac{1}{2}[\alpha, \beta]) \|$$
$$\leq c_5 \cdot (2 \cdot \|\alpha\| + 2 \cdot \|\beta\| + \|[\alpha, \beta]\|) \cdot (\|[\alpha, \beta]\|)$$
$$\leq c_5 \cdot (2 + c_2 \cdot \epsilon_4) \cdot c_2 \cdot \|\alpha\| \cdot \|\beta\| \cdot (\|\alpha\| + \|\beta\|).$$

In the last inequality we used
$$\|[\alpha, \beta]\| \leq c_2 \cdot \|\alpha\| \cdot \|\beta\| \leq c_2 \cdot \epsilon_4 \cdot \|\beta\|.$$

Therefore (2.5.2) holds with

$$c_6 := 2c_2c_3 + c_5 \cdot (2 + c_2\epsilon_4) \cdot c_2.$$

$\square$

Proof of Theorem 2.1.2. Take $c_0(G) := c_6$ and $\epsilon_0(G) := \epsilon_4$. Since $\epsilon_4 \leq \epsilon_2$ and

$$c_6 \geq c_5 \geq \max(c_4, \tfrac{3}{2}) \geq c_4 \geq c_3 \geq c_2,$$

the assertions of the theorem are contained on Lemmas 2.3.1, 2.4.2, 2.4.8 and 2.5.1. $\square$

3 Multiplicative Integration in Dimension One

Nonabelian multiplicative integration in dimension 1 is classical, dating back to the work of Volterra (cf. [DF]). In modern differential geometry it is usually viewed as the holonomy of a connection along a path. We prefer to do everything from scratch, for several reasons: this allows us to introduce notation; it serves as a warm-up for the much more difficult 2-dimensional integration; and also to cover the case of piecewise smooth differential forms.

3.1 Binary Tessellations

Recall that $\mathbf{I}^1$ is the unit line segment in $\mathbf{A}^1 = \mathbf{A}^1(\mathbb{R})$, which we view as an oriented polyhedron (cf. Chapter 1). The vertices (endpoints) of $\mathbf{I}^1$ are $v_0 = 0$ and $v_1 = 1$, and the coordinate function is t_1.

Let X be a polyhedron. Consider a linear map $\sigma : \mathbf{I}^1 \to X$. The length of the line segment $Z := \sigma(\mathbf{I}^1) \subset X$ (possibly zero) is denoted by $\mathrm{len}(\sigma)$. If σ is not constant then it is a conformal map, and the scaling factor is precisely $\mathrm{len}(\sigma)$. In this case σ determines an orientation on the 1-dimensional polyhedron Z. We may choose an orthonormal linear coordinate function f_1 on Z, such that $\mathrm{d}f_1$ is the orientation of Z. Then $\sigma^*(\mathrm{d}f_1) = \mathrm{len}(\sigma) \cdot \mathrm{d}t_1$.

Next consider a piecewise linear map $\sigma : \mathbf{I}^1 \to X$. By definition (cf. Section 1.6) there is a linear triangulation $\{Y_j\}_{j \in J}$ of $\mathbf{I}^1$ such that $\sigma|_{Y_j} : Y_j \to X$ is a linear map for every $j \in J$. We define

$$\mathrm{len}(\sigma) := \sum_{j \in J_1} \mathrm{len}(\sigma_j).$$

It is convenient to have a composition operation for piecewise linear maps. Suppose $\sigma : \mathbf{I}^1 \to X$ and $\rho : \mathbf{I}^1 \to \mathbf{I}^1$ are piecewise linear maps. Then the set-theoretical composition $\sigma \circ \rho$ is also a piecewise linear map $\mathbf{I}^1 \to X$. Given finite sequences $\sigma = (\sigma_i)_{i=1,\dots,m}$ and $\rho = (\rho_j)_{j=1,\dots,n}$ of piecewise linear maps $\sigma_i : \mathbf{I}^1 \to X$ and $\rho_j : \mathbf{I}^1 \to \mathbf{I}^1$, we define the

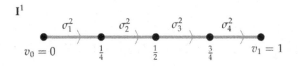

FIGURE 6. The 2-nd binary tessellation of $\mathbf{I}^1$. The arrowheads indicate the orientations of the linear maps $\sigma_i^2 : \mathbf{I}^1 \to \mathbf{I}^1$.

sequence of piecewise linear maps

$$\sigma \circ \rho := (\sigma_i \circ \rho_j)_{(i,j) \in \{1,\dots,m\} \times \{1,\dots,n\}}$$

in lexicographical order, i.e.

$$(3.1.1) \qquad \sigma \circ \rho = (\sigma_1 \circ \rho_1, \ \sigma_1 \circ \rho_2, \ \dots, \sigma_2 \circ \rho_1, \ \sigma_2 \circ \rho_2, \ \dots, \sigma_m \circ \rho_n) \ .$$

Definition 3.1.2. For any $k \geq 0$, the *k-th binary tessellation* of $\mathbf{I}^1$ is the sequence

$$\text{tes}^k \, \mathbf{I}^1 = (\sigma_1^k, \dots, \sigma_{2^k}^k)$$

of linear maps in $\sigma_i^k : \mathbf{I}^1 \to \mathbf{I}^1$ defined recursively as follows.

- For $k = 0$ we define σ_1^0 to be the identity map of $\mathbf{I}^1$.
- For $k = 1$ we take the linear maps σ_1^1, σ_2^1 defined on vertices by

$$\sigma_1^1(v_0, v_1) := (v_0, \tfrac{1}{2}),$$

$$\sigma_2^1(v_0, v_1) := (\tfrac{1}{2}, v_1).$$

- For $k \geq 1$ we define

$$\text{tes}^{k+1} \, \mathbf{I}^1 := (\text{tes}^1 \, \mathbf{I}^1) \circ (\text{tes}^k \, \mathbf{I}^1),$$

using the convention (3.1.1).

We call σ_1^0 the *basic map*.

See Figure 6 for an illustration of $\text{tes}^2 \, \mathbf{I}^1$.

3.2 Riemann Products

Let $\mathfrak{g}$ be a finite dimensional Lie algebra over $\mathbb{R}$. For any n-dimensional polyhedron X we then have the DG Lie algebra of piecewise smooth $\mathfrak{g}$-valued differential forms

$$\Omega_{\text{pws}}(X) \otimes \mathfrak{g} = \bigoplus_{p=0}^{n} \Omega_{\text{pws}}^p(X) \otimes \mathfrak{g}$$

(Section 1.6). The operations are as follows: for $\alpha_i \in \Omega_{\text{pws}}^{p_i}(X)$ and $\gamma_i \in \mathfrak{g}$ one has

$$d(\alpha_1 \otimes \gamma_1) = d(\alpha_1) \otimes \gamma_1 \in \Omega_{\text{pws}}^{p_1+1}(X) \otimes \mathfrak{g}$$

and

$$[\alpha_1 \otimes \gamma_1, \alpha_2 \otimes \gamma_2] = (\alpha_1 \wedge \alpha_2) \otimes [\gamma_1, \gamma_2] \in \Omega_{\text{pws}}^{p_1+p_2}(X) \otimes \mathfrak{g}.$$

By definition, any particular piecewise smooth differential form α belongs to $\Omega_{\text{pws}}(X; T) \otimes \mathfrak{g}$, for some linear triangulation T of X. This construction is functorial in the following sense. Suppose $f : X \to Y$ is a piecewise linear map of polyhedra, and $\phi : \mathfrak{g} \to \mathfrak{h}$ is map of Lie algebras. Then there is a homomorphism of DG Lie algebras

$$f^* \otimes \phi : \Omega_{\text{pws}}(Y) \otimes \mathfrak{g} \to \Omega_{\text{pws}}(X) \otimes \mathfrak{h}.$$

In case $\mathfrak{h} = \mathfrak{g}$ and ϕ is the identity map, we shall often write f^* instead of $f^* \otimes \phi$.

From now on in this chapter we consider a Lie group G, with Lie algebra $\mathfrak{g}$, and a polyhedron X. We fix some euclidean norm $\|-\|$ on the vector space $\mathfrak{g}$. As in Chapter 1 we also fix an open neighborhood $V_0(G)$ of 1 in G on which $\log_G$ is defined, a convergence radius $\epsilon_0(G)$, and a commutativity constant $c_0(G)$. The choices of $V_0(G)$, $\epsilon_0(G)$ and $c_0(G)$ are auxiliary only; they are needed for the proofs, but do not effect the results.

Definition 3.2.1. Let $\alpha \in \Omega_{\text{pws}}^1(\mathbf{I}^1) \otimes \mathfrak{g}$. The *basic Riemann product of α on $\mathbf{I}^1$* is the element

$$\text{RP}_0(\alpha \mid \mathbf{I}^1) \in G$$

defined as follows. Let $w := \frac{1}{2}$, namely the midpoint of $\mathbf{I}^1$.

- Suppose w is a smooth point of α. Then there is some 1-dimensional simplex Y in $\mathbf{I}^1$, such that $w \in \text{Int}\, Y$ and $\alpha|_Y$ is smooth. Let $\tilde{\alpha} \in \mathcal{O}(Y) \otimes \mathfrak{g}$ be the coefficient of $\alpha|_Y$, as in Definition 1.8.2; namely

$$\alpha|_Y = \tilde{\alpha} \cdot dt_1.$$

  We define

$$\text{RP}_0(\alpha \mid \mathbf{I}^1) := \exp_G\big(\tilde{\alpha}(w)\big).$$

- If w is a singular point of α, then we let $\text{RP}_0(\alpha \mid \mathbf{I}^1) := 1$.

Observe that the element $\tilde{\alpha}(w) \in \mathfrak{g}$ in the definition above is independent of the simplex Y.

Definition 3.2.2. Let $\alpha \in \Omega^1_{\text{pws}}(X) \otimes \mathfrak{g}$, and let $\sigma : \mathbf{I}^1 \to X$ be a piecewise linear map. Then $\sigma^*(\alpha) \in \Omega^1_{\text{pws}}(\mathbf{I}^1) \otimes \mathfrak{g}$, and we define the *basic Riemann product of α along σ* to be

$$\text{RP}_0(\alpha \,|\, \sigma) := \text{RP}_0\big(\sigma^*(\alpha) \,|\, \mathbf{I}^1\big) \in G.$$

Note that if σ is a constant map then $\sigma^*(\alpha) = 0$, so $\text{RP}_0(\alpha \,|\, \sigma) = 1$.

Definition 3.2.3. Let $\alpha \in \Omega^1_{\text{pws}}(X) \otimes \mathfrak{g}$, and let $\sigma : \mathbf{I}^1 \to X$ be a piecewise linear map. For $k \geq 0$ we define the *k-th refined Riemann product of α along σ* to be

$$\text{RP}_k(\alpha \,|\, \sigma) := \prod_{i=1}^{2^k} \text{RP}_0(\alpha \,|\, \sigma \circ \sigma_i^k) \in G,$$

using the k-th binary tessellation $\text{tes}^k \, \mathbf{I}^1 = (\sigma_1^k, \ldots, \sigma_{2^k}^k)$ and the convention (2.1.1).

3.3 Convergence of Riemann Products

As before we are given a form $\alpha \in \Omega^1_{\text{pws}}(X) \otimes \mathfrak{g}$. Recall $\|\alpha\|_{\text{Sob}}$, the Sobolev norm to order 2 of α, from Section 1.7.

Lemma 3.3.1. *There are constants $c_1(\alpha)$ and $\epsilon_1(\alpha)$ with the following properties.*

(i) *These inequalities hold:*

$$0 < \epsilon_1(\alpha) \leq 1,$$
$$c_1(\alpha) \geq 1,$$
$$\epsilon_1(\alpha) \cdot c_1(\alpha) \leq \tfrac{1}{2} \cdot \epsilon_0(G).$$

(ii) *For any piecewise linear map $\sigma : \mathbf{I}^1 \to X$ such that $\text{len}(\sigma) < \epsilon_1(\alpha)$, and for any sufficiently large k, one has*

$$\text{RP}_k(\alpha \,|\, \sigma) \in V_0(G)$$

and

$$\|\log_G\big(\text{RP}_k(\alpha \,|\, \sigma)\big)\| \leq c_1(\alpha) \cdot \text{len}(\sigma).$$

(iii) *Moreover, if the map σ in (ii) is linear, then the assertions there hold for any $k \geq 0$.*

Proof. We are given a piecewise linear map $\sigma : \mathbf{I}^1 \to X$. Let us write $\epsilon := \text{len}(\sigma)$. Excluding the trivial case, we may assume that $\epsilon > 0$.

Take $k \geq 0$. For any index $i \in \{1, \ldots, 2^k\}$ let $W_i := \sigma_i^k(\mathbf{I}^1)$ and $w_i := \sigma_i^k(\tfrac{1}{2})$; so w_i is the midpoint of the segment (1-dimensional polyhedron)

W_i. Define $\epsilon_i := \mathrm{len}(\sigma \circ \sigma_i^k)$, $Z_i := \sigma(W_i)$ and $z_i := \sigma(w_i)$. Note that $\sum_i \epsilon_i = \epsilon$.

We will say that an index i is good if the map $\sigma|_{W_i}$ is linear and injective. In this case Z_i is a segment of length ϵ_i and midpoint z_i. Otherwise we will call i a bad index. The sets of good and bad indices are denoted by $\mathrm{good}(k)$ and $\mathrm{bad}(k)$ respectively. Let m be the number of singular points of the map σ. Then $|\mathrm{bad}(k)| \leq m$. In particular, if σ is linear then $\mathrm{bad}(k) = \varnothing$.

Let

$$\alpha' := \sigma^*(\alpha) \in \Omega^1_{\mathrm{pws}}(\mathbf{I}^1) \otimes \mathfrak{g}.$$

For an index i we define an element $\lambda_i \in \mathfrak{g}$ as follows. If w_i is a smooth point of α', then let $\tilde{\alpha}'_i$ be the coefficient of α' near w_i, and let

$$\lambda_i := (\tfrac{1}{2})^k \cdot \tilde{\alpha}'_i(w_i).$$

Otherwise we let $\lambda_i := 0$. In any case we have

$$\exp_G(\lambda_i) = \mathrm{RP}_0(\alpha' \,|\, \sigma_i^k) = \mathrm{RP}_0(\alpha \,|\, \sigma \circ \sigma_i^k).$$

Therefore

(3.3.2)
$$\prod_{i=1}^{2^k} \exp_G(\lambda_i) = \mathrm{RP}_k(\alpha \,|\, \sigma).$$

Furthermore, if i is a good index and w_i is a smooth point of α', then z_i is a smooth point $\alpha|_{Z_i}$, and then

(3.3.3)
$$\lambda_i = \epsilon_i \cdot \tilde{\alpha}_i(z_i),$$

where $\tilde{\alpha}_i$ is the coefficient of $\alpha|_{Z_i}$ near z_i.

Now here are the estimates. For any index i we have

$$\|\lambda_i\| \leq (\tfrac{1}{2})^k \cdot \|\alpha'\|_{\mathrm{Sob}}.$$

If i is a good index then by (3.3.3) we have

$$\|\lambda_i\| \leq \epsilon_i \cdot \|\alpha\|_{\mathrm{Sob}}.$$

Hence

(3.3.4)
$$\sum_{i=1}^{2^k} \|\lambda_i\| = \sum_{i \in \mathrm{good}(k)} \|\lambda_i\| + \sum_{i \in \mathrm{bad}(k)} \|\lambda_i\|$$
$$\leq \epsilon \cdot \|\alpha\|_{\mathrm{Sob}} + m \cdot (\tfrac{1}{2})^k \cdot \|\alpha'\|_{\mathrm{Sob}}.$$

Let

$$\epsilon_1(\alpha) := \tfrac{1}{2} \cdot (1 + \|\alpha\|_{\mathrm{Sob}})^{-1} \cdot \epsilon_0(G)$$

and
$$c_1(\alpha) := 2 \cdot \left(1 + c_0(G) \cdot \|\alpha\|_{\text{Sob}}\right).$$
Choose k_0 large enough so that
$$m \cdot \left(\tfrac{1}{2}\right)^{k_0} \cdot \|\alpha'\|_{\text{Sob}} < \min\left(\tfrac{1}{2} \cdot \epsilon_0(G), \ \tfrac{1}{2} \cdot c_1(\alpha) \cdot \epsilon \cdot (1 + c_0(G))^{-1}\right).$$
If σ is linear then $m = 0$, and we may take $k_0 := 0$.

Now suppose that $\epsilon < \epsilon_1(\alpha)$ and $k \geq k_0$. According to inequality (3.3.4) we have
$$\sum_{i=1}^{2^k} \|\lambda_i\| < \epsilon_0(G).$$
Therefore by property (ii) of Theorem 2.1.2 and by formula (3.3.2) we get
$$\mathrm{RP}_k(\alpha \,|\, \sigma) \in V_0(G)$$
and
$$\|\log_G\left(\mathrm{RP}_k(\alpha \,|\, \sigma)\right)\| \leq c_0(G) \cdot \left(\sum_{i=1}^{2^k} \|\lambda_i\|\right) \leq c_1(\alpha) \cdot \epsilon.$$
$$\square$$

Remark 3.3.5. Heuristically we think of ϵ in the proof of the lemma above as a "tiny" size. In the "tiny scale" we can measure things (i.e. $\|\log_G(g)\|$ is defined), and we can use Taylor series and CBH series.

Lemma 3.3.6. *There are constants $c_2(\alpha)$ and $\epsilon_2(\alpha)$ with the following properties.*

(i) *These inequalities hold:*
$$0 < \epsilon_2(\alpha) \leq \epsilon_1(\alpha),$$
$$c_2(\alpha) \geq c_1(\alpha).$$

(ii) *Suppose $\sigma : \mathbf{I}^1 \to X$ is a linear map such that $\mathrm{len}(\sigma) < \epsilon_2(\alpha)$, and $\alpha|_{\sigma(\mathbf{I}^1)}$ is smooth. Then for any $k \geq 0$ one has*
$$\|\log_G\left(\mathrm{RP}_k(\alpha \,|\, \sigma)\right) - \log_G\left(\mathrm{RP}_0(\alpha \,|\, \sigma)\right)\| \leq c_2(\alpha) \cdot \mathrm{len}(\sigma)^3.$$

Proof. Take
$$\epsilon_2(\alpha) := \min\left(\epsilon_1(\alpha), \tfrac{1}{4}\epsilon_0(G) \cdot (1 + \|\alpha\|_{\text{Sob}})^{-1}\right).$$

Let σ be a linear map with $\epsilon := \mathrm{len}(\alpha)$ satisfying $0 < \epsilon < \epsilon_2(\alpha)$. Let $Z := \sigma(\mathbf{I}^1)$, which is a 1-dimensional oriented polyhedron, and $z := \sigma(\tfrac{1}{2})$, the midpoint of Z. Choose a positively oriented orthonormal linear coordinate function s_1 on Z, such that $s_1(z) = 0$.

Let $\tilde{\alpha} \in \mathcal{O}(Z) \otimes \mathfrak{g}$ be the coefficient of $\alpha|_Z$, i.e. $\alpha|_Z = \tilde{\alpha} \cdot dt_1$. Consider the Taylor expansion to second order of the smooth function $\tilde{\alpha} : Z \to \mathfrak{g}$, around the point z:

$$(3.3.7) \qquad \tilde{\alpha}(x) = a_0 + s_1(x) \cdot a_1 + s_1(x)^2 \cdot g(x)$$

for $x \in Z$, where $a_0 := \tilde{\alpha}(z) \in \mathfrak{g}$, $a_1 := (\frac{\partial}{\partial s_1}\tilde{\alpha})(z) \in \mathfrak{g}$, and $g : Z \to \mathfrak{g}$ is a continuous function. We know that

$$\|a_0\|, \|a_1\|, \|g(x)\| \leq \|\alpha\|_{\mathrm{Sob}} .$$

And, as we have seen before,

$$(3.3.8) \qquad \mathrm{RP}_0(\alpha \,|\, \sigma) = \exp_G(\epsilon a_0).$$

Take $k \geq 0$. For $i \in \{1, \ldots, 2^k\}$ let $\sigma_i := \sigma \circ \sigma_i^k$ and $z_i := \sigma_i(\frac{1}{2})$. Since $\epsilon < \epsilon_2(\alpha)$ it follows that

$$(3.3.9) \qquad \sum_{i=1}^{2^k} \|(\tfrac{1}{2})^k \epsilon \cdot a_0\| \leq \epsilon \cdot \|\alpha\|_{\mathrm{Sob}} \leq \tfrac{1}{4}\epsilon_0(G) .$$

Because $|s_1(z_i)| \leq \frac{1}{2}\epsilon$ we also have

$$(3.3.10) \qquad \sum_{i=1}^{2^k} \|(\tfrac{1}{2})^k \epsilon \cdot s_1(z_i) \cdot a_1\| \leq \tfrac{1}{2}\epsilon^2 \cdot \|\alpha\|_{\mathrm{Sob}} \leq \tfrac{1}{4}\epsilon_0(G)$$

and

$$(3.3.11) \qquad \sum_{i=1}^{2^k} \|(\tfrac{1}{2})^k \epsilon \cdot s_1(z_i)^2 \cdot g(z_i)\| \leq \tfrac{1}{4}\epsilon^3 \cdot \|\alpha\|_{\mathrm{Sob}} \leq \tfrac{1}{4}\epsilon_0(G) .$$

Define

$$\lambda_i := (\tfrac{1}{2})^k \epsilon \cdot \tilde{\alpha}(z_i) \in \mathfrak{g}.$$

By the Taylor expansion (3.3.7) we have

$$(3.3.12) \qquad \lambda_i = (\tfrac{1}{2})^k \epsilon \cdot a_0 + (\tfrac{1}{2})^k \epsilon \cdot s_1(z_i) \cdot a_1 + (\tfrac{1}{2})^k \epsilon \cdot s_1(z_i)^2 \cdot g(z_i).$$

Since $\sigma^*(ds_1) = \epsilon \cdot dt_1$, we see that

$$\mathrm{RP}_0(\alpha \,|\, \sigma_i) = \exp_G(\lambda_i),$$

and hence

$$(3.3.13) \qquad \mathrm{RP}_k(\alpha \,|\, \sigma) = \prod_{i=1}^{2^k} \exp_G(\lambda_i).$$

Because the constant terms in the Taylor expansions (3.3.12) of the λ_i are all equal to $(\frac{1}{2})^k \epsilon \cdot a_0$, we can use property (v) of Theorem 2.1.2, together with the estimates (3.3.9), (3.3.10), (3.3.11), to deduce
(3.3.14)

$$\left\| \log_G \left(\prod_{i=1}^{2^k} \exp_G(\lambda_i) \right) - \sum_{i=1}^{2^k} \lambda_i \right\|$$
$$\leq c_0(G) \cdot \left(\epsilon \cdot \|\alpha\|_{\mathrm{Sob}} \cdot (1 + \tfrac{1}{2}\epsilon + \tfrac{1}{4}\epsilon^2) \right) \cdot \left(\epsilon \cdot \|\alpha\|_{\mathrm{Sob}} \cdot (\tfrac{1}{2}\epsilon + \tfrac{1}{4}\epsilon^2) \right)$$
$$\leq \epsilon^3 \cdot c_0(G) \cdot \|\alpha\|_{\mathrm{Sob}}^2 \cdot 2.$$

Trivially the sum of the constant terms of the Taylor expansions of the λ_i is

$$\sum_{i=1}^{2^k} (\tfrac{1}{2})^k \epsilon \cdot a_0 = \epsilon \cdot a_0.$$

The linear terms satisfy

$$s_1(z_i) = -s_1(z_{2^k - i})$$

because of symmetry; and therefore they cancel out:

$$\sum_{i=1}^{2^k} (\tfrac{1}{2})^k \epsilon \cdot s_1(z_i) \cdot a_1 = 0.$$

Therefore, using the estimate (3.3.11) to eliminate the quadratic terms of the Taylor expansions, we conclude that

(3.3.15) $$\left\| \sum_{i=1}^{2^k} \lambda_i - \epsilon a_0 \right\| \leq \tfrac{1}{4}\epsilon^3 \cdot \|\alpha\|_{\mathrm{Sob}}.$$

Finally we define

$$c_2(\alpha) := \max\left(c_1(\alpha), 2c_0(G) \cdot \|\alpha\|_{\mathrm{Sob}}^2 + \tfrac{1}{4} \cdot \|\alpha\|_{\mathrm{Sob}} \right).$$

Combining equations (3.3.8) and (3.3.13), plus the estimates (3.3.15) and (3.3.14), we obtain

$$\left\| \log_G \left(\mathrm{RP}_0(\alpha \mid \sigma) \right) - \log_G \left(\mathrm{RP}_k(\alpha \mid \sigma) \right) \right\| \leq c_2(\alpha) \cdot \epsilon^3.$$

$\square$

Definition 3.3.16. Let us fix a constant $\epsilon_2(\alpha)$ as in Lemma 3.3.6. A piecewise linear map $\sigma : \mathbf{I}^1 \to X$ with $\mathrm{len}(\sigma) < \epsilon_2(\alpha)$ will be called an *α-tiny piecewise linear map* (in this chapter).

Remark 3.3.17. We shall use the term "tiny" several times in the book, each time with a new meaning, depending on context. The notion "tiny" should be considered as "local to each chapter".

Lemma 3.3.18. *Let $\sigma : I^1 \to X$ be an α-tiny linear map. Then there is a constant $c_3(\alpha, \sigma) \geq 0$, such that for any integers $k' \geq k \geq 0$ one has*

$$\left\| \log_G \left(RP_{k'}(\alpha \mid \sigma) \right) - \log_G \left(RP_k(\alpha \mid \sigma) \right) \right\| \leq c_3(\alpha, \sigma) \cdot \mathrm{len}(\sigma) \cdot (\tfrac{1}{2})^k.$$

Proof. We may assume that $\mathrm{len}(\sigma) > 0$. Let $Z := \sigma(I^1) \subset X$, and let m be the number of singular points of the differential form $\alpha|_Z$.

Take $k \geq 0$. For an index $i \in \{1, \ldots, 2^k\}$ define $\sigma_i := \sigma \circ \sigma_i^k$ and $Z_i := \sigma_i(I^1)$. We say that i is good if $\alpha|_{Z_i}$ is smooth, and otherwise i is bad. The set of good and bad indices are denoted by $\mathrm{good}(k)$ and $\mathrm{bad}(k)$ respectively. Since for any singular point x of $\alpha|_Z$ there is at most one index i such that $x \in \mathrm{Int}\, Z_i$, it follows that the cardinality of $\mathrm{bad}(k)$ is at most m.

Next take $k' \geq k \geq 0$, and let $l := k' - k$. By the recursive definition of the binary tessellations we have

$$RP_k(\alpha \mid \sigma) = \prod_{i=1}^{2^k} RP_0(\alpha \mid \sigma_i)$$

and

$$RP_{k'}(\alpha \mid \sigma) = \prod_{i=1}^{2^k} RP_l(\alpha \mid \sigma_i).$$

If $i \in \mathrm{good}(k)$ then by Lemma 3.3.6 we know that

$$\left\| \log_G \left(RP_l(\alpha \mid \sigma_i) \right) - \log_G \left(RP_0(\alpha \mid \sigma)_i \right) \right\| \leq c_2(\alpha) \cdot \mathrm{len}(\sigma)^3 \cdot (\tfrac{1}{2})^{3k}.$$

On the other hand, $i \in \mathrm{bad}(k)$, then by Lemma 3.3.1 we know that

$$\left\| \log_G \left(RP_l(\alpha \mid \sigma_i) \right) - \log_G \left(RP_0(\alpha \mid \sigma)_i \right) \right\| \leq 2 \cdot c_1(\alpha) \cdot \mathrm{len}(\sigma) \cdot (\tfrac{1}{2})^k.$$

Therefore by property (iv) of Theorem 2.1.2 we have

$$\left\| \log_G \left(RP_{k'}(\alpha \mid \sigma) \right) - \log_G \left(RP_k(\alpha \mid \sigma) \right) \right\|$$
$$= \left\| \log_G \left(\prod_{i=1}^{2^k} RP_l(\alpha \mid \sigma_i) \right) - \log_G \left(RP_0(\alpha \mid \sigma_i) \right) \right\|$$
$$\leq c_0(G) \cdot \left(|\mathrm{bad}(k)| \cdot 2 \cdot c_1(\alpha) \cdot \mathrm{len}(\sigma) \cdot (\tfrac{1}{2})^k \right.$$
$$\left. + |\mathrm{good}(k)| \cdot c_2(\alpha) \cdot \mathrm{len}(\sigma)^3 \cdot (\tfrac{1}{2})^{3k} \right).$$

Thus we may take

$$c_3(\alpha, \sigma) := c_0(G) \cdot \left(2m \cdot c_1(\alpha) + c_2(\alpha) \right).$$

$\square$

Theorem 3.3.19. *Let X be a polyhedron, let $\alpha \in \Omega^1_{\mathrm{pws}}(X) \otimes \mathfrak{g}$, and let $\sigma : I^1 \to X$ be a piecewise linear map. Then the limit $\lim_{k \to \infty} RP_k(\alpha \mid \sigma)$ exists.*

Proof. For any k we have

$$\mathrm{RP}_k(\alpha \,|\, \sigma) = \prod_{i=1}^{2^k} \mathrm{RP}_0(\alpha \,|\, \sigma \circ \sigma_i^k) = \prod_{i=1}^{2^k} \mathrm{RP}_0\big((\sigma \circ \sigma_i^k)^*(\alpha) \,|\, \mathbf{I}^1\big)$$

$$= \prod_{i=1}^{2^k} \mathrm{RP}_0\big(\sigma^*(\alpha) \,|\, \sigma_i^k\big) = \mathrm{RP}_k\big(\sigma^*(\alpha) \,|\, \sigma_1^0\big)$$

by definition. So after replacing α with $\sigma^*(\alpha)$, we can assume that $X = \mathbf{I}^1$, and we have to prove that the limit $\lim_{k \to \infty} \mathrm{RP}_k(\alpha \,|\, \sigma_1^0)$ exists.

Take k large enough such that for each $i \in \{1, \ldots, 2^k\}$ the linear map σ_i^k is α-tiny. For any $k' \geq 0$ we have

$$\mathrm{RP}_{k+k'}(\alpha \,|\, \sigma_1^0) = \prod_{i=1}^{2^k} \mathrm{RP}_{k'}(\alpha \,|\, \sigma_i^k).$$

Thus it suffices to prove that for any i the limit $\lim_{k' \to \infty} \mathrm{RP}_{k'}(\alpha \,|\, \sigma_i^k)$ exists.

We have now reduced our problem to showing that for any α-tiny linear map $\sigma : \mathbf{I}^1 \to X$ the limit $\lim_{k \to \infty} \mathrm{RP}_k(\alpha \,|\, \sigma)$ exists. But this follows immediately from Lemma 3.3.18. $\qquad\square$

Definition 3.3.20. In the situation of Theorem 3.3.19, the *multiplicative integral of α on σ* is

$$\mathrm{MI}(\alpha \,|\, \sigma) := \lim_{k \to \infty} \mathrm{RP}_k(\alpha \,|\, \sigma) \in G.$$

If $X = \mathbf{I}^1$ and $\sigma = \sigma_1^0$, the basic map, then we write $\mathrm{MI}(\alpha \,|\, \mathbf{I}^1) := \mathrm{MI}(\alpha \,|\, \sigma_1^0)$.

Remark 3.3.21. If the group G is *abelian*, then $\mathrm{RP}_k(\alpha \,|\, \sigma)$ is the exponential of a Riemann sum, and therefore in the limit we get

$$\mathrm{MI}(\alpha \,|\, \sigma) = \exp_G\Big(\int_\sigma \alpha\Big).$$

Proposition 3.3.22. *Let $\sigma : \mathbf{I}^1 \to X$ be a piecewise linear map.*

(1) *If σ is α-tiny (see Definition 3.3.16), then $\mathrm{MI}(\alpha \,|\, \sigma) \in V_0(G)$, and*

$$\big\| \log_G\big(\mathrm{MI}(\alpha \,|\, \sigma)\big) \big\| \leq c_1(\alpha) \cdot \mathrm{len}(\sigma).$$

(2) *If σ is linear and α-tiny, and if $\alpha|_{\sigma(\mathbf{I}^1)}$ is smooth, then for any $k \geq 0$ one has*

$$\big\| \log_G\big(\mathrm{MI}(\alpha \,|\, \sigma)\big) - \log_G\big(\mathrm{RP}_0(\alpha \,|\, \sigma)\big) \big\| \leq c_2(\alpha) \cdot \mathrm{len}(\sigma)^3.$$

(3) *For any $k \geq 0$ one has*

$$\mathrm{MI}(\alpha \,|\, \sigma) = \prod_{i=1}^{2^k} \mathrm{MI}(\alpha \,|\, \sigma \circ \sigma_i^k).$$

Proof. (1) By Lemma 3.3.1, for sufficiently large k we have $\mathrm{RP}_k(\alpha \,|\, \sigma) \in V_0(G)$ and

$$\left\| \log_G \big(\mathrm{RP}_k(\alpha \,|\, \sigma) \big) \right\| \leq c_1(\alpha) \cdot \mathrm{len}(\sigma) \leq \tfrac{1}{2} \cdot \epsilon_0(G) \,.$$

Let B be the closed ball of radius $\tfrac{1}{2}\epsilon_0(G)$ in $\mathfrak{g}$, and let $Z := \exp_G(B)$, which is a compact subset of G. Since for every k one has $\mathrm{RP}_k(\alpha \,|\, \sigma) \in Z$, it follows that in the limit $\mathrm{MI}(\alpha \,|\, \sigma) \in Z \subset V_0(G)$. The bound on $\left\| \log_G \big(\mathrm{MI}(\alpha \,|\, \sigma) \big) \right\|$ is then obvious.

(2) This is immediate from Lemma 3.3.6.

(3) For every $k' \geq 0$ we have by definition

$$\mathrm{RP}_{k+k'}(\alpha \,|\, \sigma) = \prod_{i=1}^{2^k} \mathrm{RP}_{k'}(\alpha \,|\, \sigma \circ \sigma_i^k).$$

Now pass to the limit $k' \to \infty$. $\qquad\square$

3.4   Functoriality of the MI

The next results are on the functoriality of the multiplicative integral with respect to G and X.

Proposition 3.4.1. *Let $\Phi : G \to H$ be a map of Lie groups, with induced Lie algebra map $\phi := \mathrm{Lie}(\Phi) : \mathfrak{g} \to \mathfrak{h}$. Let $f : Y \to X$ be a piecewise linear map between polyhedra, and let $\alpha \in \Omega^1_{\mathrm{pws}}(X) \otimes \mathfrak{g}$. Then for any piecewise linear map $\sigma : \mathbf{I}^1 \to Y$ one has*

$$\Phi\big(\mathrm{MI}(\alpha \,|\, f \circ \sigma) \big) = \mathrm{MI}\big((f^* \otimes \phi)(\alpha) \,|\, \sigma \big)$$

in H.

Proof. It suffices to consider f and Φ separately; so we look at two cases.

Case 1. $H = G$ and Φ is the identity map. Here for every $k \geq 0$ and $i \in \{1, \dots, 2^k\}$ we have

$$\mathrm{RP}_0(\alpha \,|\, f \circ \sigma \circ \sigma_i^k)) = \mathrm{RP}_0\big((f \circ \sigma)^*(\alpha) \,|\, \sigma_i^k \big) = \mathrm{RP}_0\big(f^*(\alpha) \,|\, \sigma \circ \sigma_i^k \big).$$

Hence

$$\mathrm{RP}_k(\alpha \,|\, f \circ \sigma) = \mathrm{RP}_k\big((f \circ \sigma)^*(\alpha) \,|\, \mathbf{I}^1 \big) = \mathrm{RP}_k\big(f^*(\alpha) \,|\, \sigma \big).$$

Going to the limit in k we see that

$$\mathrm{MI}(\alpha \,|\, f \circ \sigma) = \mathrm{MI}((f \circ \sigma)^*(\alpha) \,|\, \mathrm{I}^1) = \mathrm{MI}(f^*(\alpha) \,|\, \sigma).$$

Case 2. Here we assume that $Y = X$ and f is the identity map. Since

$$\mathrm{MI}(\alpha \,|\, \sigma) = \mathrm{MI}(\sigma^*(\alpha) \,|\, \sigma_1^0)$$

and

$$\mathrm{MI}(\phi(\alpha) \,|\, \sigma) = \mathrm{MI}((\sigma^* \otimes \phi)(\alpha) \,|\, \sigma_1^0) = \mathrm{MI}(\phi(\sigma^*(\alpha)) \,|\, \sigma_1^0),$$

we can replace σ with σ_1^0 and α with $\sigma^*(\alpha)$. Therefore we can assume that σ is a linear map.

Put a euclidean norm on $\mathfrak{h}$ such that $\phi : \mathfrak{g} \to \mathfrak{h}$ has operator norm $\|\phi\| \leq 1$. This implies that $\|\phi(\alpha)\|_{\mathrm{Sob}} \leq \|\alpha\|_{\mathrm{Sob}}$. So we can assume that $\epsilon_1(\phi(\alpha)) \geq \epsilon_1(\alpha)$ and $c_1(\phi(\alpha)) \leq c_1(\alpha)$ (cf. proof of Lemma 3.3.1).

Take k large enough such that $\sigma \circ \sigma_i^k$ is α-tiny for every $i \in \{1, \ldots, 2^k\}$. Then $\sigma \circ \sigma_i^k$ is also $\phi(\alpha)$-tiny.

By part (3) of Proposition 3.3.22 we have

$$\mathrm{MI}(\alpha \,|\, \sigma) = \prod_{i=1}^{2^k} \mathrm{MI}(\alpha \,|\, \sigma \circ \sigma_i^k)$$

and

$$\mathrm{MI}(\phi(\alpha) \,|\, \sigma) = \prod_{i=1}^{2^k} \mathrm{MI}(\phi(\alpha) \,|\, \sigma \circ \sigma_i^k).$$

So it suffices to prove that

$$\Phi\big(\mathrm{MI}(\alpha \,|\, \sigma \circ \sigma_i^k)\big) = \mathrm{MI}(\phi(\alpha) \,|\, \sigma \circ \sigma_i^k)$$

for every i.

By this reduction we can assume that σ is α-tiny and also $\phi(\alpha)$-tiny. Take any $k \geq 0$, and for every $i \in \{1, \ldots, 2^k\}$ let $w_i := \sigma_i^k(\tfrac{1}{2}) \in \mathrm{I}^1$. If w_i is a smooth point of

$$\sigma^*(\alpha) \in \Omega_{\mathrm{pws}}^1(\mathrm{I}^1) \otimes \mathfrak{g},$$

then it is also a smooth point of

$$(\sigma^* \otimes \phi)(\alpha) \in \Omega_{\mathrm{pws}}^1(\mathrm{I}^1) \otimes \mathfrak{h}.$$

In this case we have

(3.4.2) $$\Phi\big(\mathrm{RP}_0(\alpha \,|\, \sigma \circ \sigma_i^k)\big) = \mathrm{RP}_0(\phi(\alpha) \,|\, \sigma \circ \sigma_i^k).$$

In case w_i is a singular point of $(\sigma^* \otimes \phi)(\alpha)$, then it is also a singular point of $\sigma^*(\alpha)$. Hence (3.4.2) also holds (both sides are 1).

The only problem is when w_i is a smooth point of $(\sigma^* \otimes \phi)(\alpha)$ but a singular point of $\sigma^*(\alpha)$.

Now we use the estimate provided by Lemma 3.3.1 (noting that $\phi \circ \log_G = \log_H \circ \Phi$):

$$\| \log_H(\Phi(\mathrm{RP}_0(\alpha \,|\, \sigma \circ \sigma_i^k))) - \log_H(\mathrm{RP}_0(\phi(\alpha) \,|\, \sigma \circ \sigma_i^k)) \|$$
$$\leq 2 \cdot c_1(\alpha) \cdot \mathrm{len}(\sigma) \cdot (\tfrac{1}{2})^k.$$

Let m be the number of singular points of $\sigma^*(\alpha)$. Then using property (iv) of Theorem 2.1.2 we get

$$\| \log_H(\Phi(\mathrm{RP}_k(\alpha \,|\, \sigma))) - \log_H(\mathrm{RP}_k(\phi(\alpha) \,|\, \sigma)) \|$$
$$\leq 2m \cdot c_1(\alpha) \cdot \mathrm{len}(\sigma) \cdot (\tfrac{1}{2})^k.$$

Since m is independent of k, in the limit $k \to \infty$ we get

$$\Phi(\mathrm{MI}(\alpha \,|\, \sigma)) = \mathrm{MI}(\phi(\alpha) \,|\, \sigma).$$

$\square$

A particular case of the above is when we are given a representation of G, namely a map of Lie groups $\Phi : G \to \mathrm{GL}_m(\mathbb{R})$ for some m. The Lie algebra of $\mathrm{GL}_m(\mathbb{R})$ is $\mathfrak{gl}_m(\mathbb{R}) = \mathrm{M}_m(\mathbb{R})$, the algebra of $m \times m$ matrices. For a matrix $a \in \mathrm{M}_m(\mathbb{R})$ let us denote by $\|a\|$ its operator norm, as linear operator $a : \mathbb{R}^m \to \mathbb{R}^m$, where $\mathbb{R}^m$ has the standard euclidean inner product.

Proposition 3.4.3. *Let* $\Phi : G \to \mathrm{GL}_m(\mathbb{R})$ *be a representation, and let* $\alpha \in \Omega^1_{\mathrm{pws}}(X) \otimes \mathfrak{g}$. *Then there is a constant* $c_4(\alpha, \Phi) \geq 0$ *such that the following holds:*

(∗) *Given a piecewise linear map* $\sigma : \mathbf{I}^1 \to X$, *let* $g := \mathrm{MI}(\alpha \,|\, \sigma) \in G$. *Then*

$$\|\Phi(g)\| \leq \exp(c_4(\alpha, \Phi) \cdot \mathrm{len}(\sigma)),$$

where exp *the usual real exponential function.*

Proof. Let us write $H := \mathrm{GL}_m(\mathbb{R})$, $\mathfrak{h} := \mathfrak{gl}_m(\mathbb{R})$ and $\alpha' := \mathrm{Lie}(\Phi)(\alpha) \in \Omega^1_{\mathrm{pws}}(X) \otimes \mathfrak{h}$. We use the operator norm on $\mathfrak{h}$ to determine the constants $\epsilon_1(\alpha')$ and $c_1(\alpha')$ in Definition 3.3.16.

Take k large enough so that $\sigma \circ \sigma_i^k$ is α'-tiny for every $i \in \{1, \ldots, 2^k\}$. Let $d := \mathrm{len}(\sigma)$ and $d_i := \mathrm{len}(\sigma \circ \sigma_i)$, so $d = \sum_{i=1}^{2^k} d_i$. Define $h_i := \mathrm{MI}(\alpha' \,|\, \sigma \circ \sigma_i^k) \in H$. Then $h_i \in V_0(H)$, and we can define $\gamma_i := \log_H(h_i) \in \mathfrak{h}$.

Since the matrix γ_i satisfies

$$\|\gamma_i\| \leq c_1(\alpha') \cdot d_i,$$

it follows that its exponential h_i satisfies

$$\|h_i\| \leq \exp(c_1(\alpha') \cdot d_i).$$

Now

$$\Phi(g) = \mathrm{MI}(\alpha' \mid \sigma) = \prod_{i=1}^{2^k} h_i,$$

so we obtain

$$\|\Phi(g)\| \leq \exp(c_1(\alpha') \cdot d).$$

Therefore we can take $c_4(\alpha, \Phi) := c_1(\alpha')$. $\qquad\square$

3.5 Strings

Definition 3.5.1. Let X be a polyhedron. A *string* in X is a sequence $\sigma = (\sigma_1, \ldots, \sigma_m)$ of piecewise linear maps $\sigma_i : \mathbf{I}^1 \to X$, such that $\sigma_i(v_1) = \sigma_{i+1}(v_0)$ for all i. The maps $\sigma_i : \mathbf{I}^1 \to X$ are called the *pieces* of σ. We write $\sigma(v_0) := \sigma_1(v_0)$ and $\sigma(v_1) := \sigma_m(v_1)$, and call these points the *initial* and *terminal* points of σ, respectively. The *length* of σ is $\mathrm{len}(\sigma) := \sum_{i=1}^{m} \mathrm{len}(\sigma_i)$.

Suppose $\alpha \in \Omega^1_{\mathrm{pws}}(X) \otimes \mathfrak{g}$. We say that σ is an α-*tiny string* if $\mathrm{len}(\sigma) < \epsilon_1(\alpha)$; cf. Definition 3.3.16.

Here are a few operations on strings. Suppose $\sigma = (\sigma_1, \ldots, \sigma_m)$ and $\tau = (\tau_1, \ldots, \tau_l)$ are two strings in X, with $\tau(v_0) = \sigma(v_1)$. Then we define the *concatenated string*

$$(3.5.2) \qquad \sigma * \tau := (\sigma_1, \ldots, \sigma_m, \tau_1, \ldots, \tau_l).$$

See Figure 7 for an illustration.

The *flip* of $\mathbf{I}^1$ is the linear bijection $\mathrm{flip} : \mathbf{I}^1 \to \mathbf{I}^1$ defined on vertices by

$$\mathrm{flip}(v_0, v_1) := (v_1, v_0).$$

Given a piecewise linear map $\sigma : \mathbf{I}^1 \to X$ we let

$$(3.5.3) \qquad \sigma^{-1} := \sigma \circ \mathrm{flip} : \mathbf{I}^1 \to X.$$

For a string $\sigma = (\sigma_1, \ldots, \sigma_m)$ in X we define the *inverse string*

$$(3.5.4) \qquad \sigma^{-1} := (\sigma_m^{-1}, \ldots, \sigma_1^{-1}).$$

The empty string is the unique string of length 0, and we denote it by $\varnothing$. For any string σ we let $\sigma * \varnothing := \sigma$ and $\varnothing * \sigma := \sigma$.

Let $f : X \to Y$ be a piecewise linear map between polyhedra, and let $\sigma = (\sigma_1, \ldots, \sigma_m)$ be a string in X. We define the string $f \circ \sigma$ in Y to be

$$(3.5.5) \qquad f \circ \sigma := (f \circ \sigma_1, \ldots, f \circ \sigma_m).$$

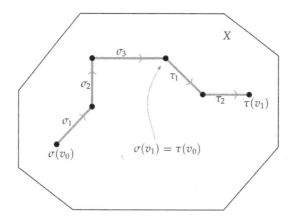

FIGURE 7. The strings $\sigma = (\sigma_1, \sigma_2, \sigma_3)$, $\tau = (\tau_1, \tau_2)$ and $\sigma * \tau$ in the polyhedron X.

Remark 3.5.6. The reason for working with strings (rather than with paths, as is the custom in algebraic topology) is that composition of strings, as defined above, is associative, whereas composition of paths is only associative up to homotopy.

It will be convenient to integrate along a string. As before G is a Lie group with Lie algebra $\mathfrak{g}$.

Definition 3.5.7. Suppose $\alpha \in \Omega^1_{\mathrm{pws}}(X) \otimes \mathfrak{g}$ and $\sigma = (\sigma_1, \ldots, \sigma_m)$ is a string in X. The multiplicative integral $\mathrm{MI}(\alpha \,|\, \sigma)$ of α on σ is

$$\mathrm{MI}(\alpha \,|\, \sigma) := \prod_{i=1}^{m} \mathrm{MI}(\alpha \,|\, \sigma_i) \in G.$$

Proposition 3.5.8. Let $\alpha \in \Omega^1_{\mathrm{pws}}(X) \otimes \mathfrak{g}$.

(1) Given strings σ and τ in X such that $\tau(v_0) = \sigma(v_1)$, one has

$$\mathrm{MI}(\alpha \,|\, \sigma * \tau) = \mathrm{MI}(\alpha \,|\, \sigma) \cdot \mathrm{MI}(\alpha \,|\, \tau).$$

(2) Given a string σ in X, one has

$$\mathrm{MI}(\alpha \,|\, \sigma^{-1}) = \mathrm{MI}(\alpha \,|\, \sigma)^{-1}.$$

(3) If σ is an α-tiny string in X (i.e $\mathrm{len}(\sigma) < \epsilon_1(\alpha)$), then $\mathrm{MI}(\alpha \,|\, \sigma) \in V_0(G)$, and

$$\left\| \log_G \left(\mathrm{MI}(\alpha \,|\, \sigma) \right) \right\| \leq c_1(\alpha) \cdot \mathrm{len}(\sigma).$$

Proof. (1) This is trivial.

(2) By part (1) it suffices to consider a piecewise linear map $\sigma : \mathbf{I}^1 \to X$. Since the flip reverses orientation on $\mathbf{I}^1$, it follows that

$$\mathrm{RP}_0(\alpha \,|\, \sigma^{-1}) = \mathrm{RP}_0(\alpha \,|\, \sigma)^{-1}.$$

From this, and the symmetry of the binary tessellations, it follows that for every $k \geq 0$ one has

$$\mathrm{RP}_k(\alpha \,|\, \sigma^{-1}) = \mathrm{RP}_k(\alpha \,|\, \sigma)^{-1}.$$

In the limit we get $\mathrm{MI}(\alpha \,|\, \sigma^{-1}) = \mathrm{MI}(\alpha \,|\, \sigma)^{-1}$.

(3) Say $\sigma = (\sigma_1, \ldots, \sigma_m)$. Take k large enough so that $2^k \geq m$. Let $\sigma' : \mathbf{I}^1 \to X$ be the unique piecewise linear map satisfying $\sigma' \circ \sigma_i^k = \sigma_i$ for $i \leq m$, and $\sigma' \circ \sigma_i^k$ is the constant map $\sigma(v_1)$ for $i > m$. Note that $\mathrm{len}(\sigma') = \mathrm{len}(\sigma)$. By Definition 3.5.7 and Proposition 3.3.22(3) we have $\mathrm{MI}(\alpha \,|\, \sigma') = \mathrm{MI}(\alpha \,|\, \sigma)$. Now use Proposition 3.3.22(1). $\qquad\square$

Proposition 3.5.9. *Let $\Phi : G \to H$ be a map of Lie groups, let $f : Y \to X$ be a piecewise linear map between polyhedra, and let $\alpha \in \Omega^1_{\mathrm{pws}}(X) \otimes \mathfrak{g}$. Then for any string σ in Y one has*

$$\Phi\big(\mathrm{MI}(\alpha \,|\, f \circ \sigma)\big) = \mathrm{MI}\big((f^* \otimes \mathrm{Lie}(\Phi))(\alpha) \,|\, \sigma\big) \in H.$$

Proof. This is an immediate consequence of Proposition 3.5.8(1) and Proposition 3.4.1. $\qquad\square$

Proposition 3.5.10. *Let $\Phi : G \to \mathrm{GL}_m(\mathbb{R})$ be a representation, and let $\alpha \in \Omega^1_{\mathrm{pws}}(X) \otimes \mathfrak{g}$. Given a string σ in X let $g := \mathrm{MI}(\alpha \,|\, \sigma)$. Then the norm of the operator $\Phi(g)$ on $\mathbb{R}^m$ satisfies*

$$\|\Phi(g)\| \leq \exp\big(c_4(\alpha, \Phi) \cdot \mathrm{len}(\sigma)\big),$$

where $c_4(\alpha, \Phi)$ is the constant from Proposition 3.4.3.

Proof. By part (1) of Proposition 3.5.8 the left side of this inequality is multiplicative with respect to the operation $*$. And clearly the right side is also multiplicative with respect to $*$. So it suffices to consider a piecewise linear map $\sigma : \mathbf{I}^1 \to X$. Now we can use Proposition 3.4.3. $\qquad\square$

For a string $\sigma = (\sigma_1, \ldots, \sigma_m)$ and a form $\alpha \in \Omega^1_{\mathrm{pws}}(X) \otimes \mathfrak{g}$ we write

$$\int_\sigma \alpha := \sum_{i=1}^m \int_{\sigma_i} \alpha \in \mathfrak{g},$$

where

$$\int_{\sigma_i} \alpha = \int_{\mathbf{I}^1} \sigma_i^*(\alpha)$$

is the usual integral of this $\mathfrak{g}$-valued piecewise smooth differential form.

Proposition 3.5.11. *Let $\alpha \in \Omega^1_{\mathrm{pws}}(X) \otimes \mathfrak{g}$, and let σ be an α-tiny string in X. Then*

$$\| \log_G (\mathrm{MI}(\alpha \,|\, \sigma)) - \int_\sigma \alpha \| \leq c_0(G) \cdot c_1(\alpha)^2 \cdot \mathrm{len}(\sigma)^2 \,.$$

What this result says, is that in the tiny scale the nonabelian integral is very close to the abelian integral.

Proof. Step 1. We begin the proof with a reduction to the case $m = 1$, and $\sigma = \sigma_1$ is a single linear map $\mathbf{I}^1 \to X$. First we append a few empty strings at the end of σ, so that the number of linear pieces becomes $m = 2^k$ for some k. This does not change $\mathrm{len}(\sigma)$ nor $\mathrm{MI}(\alpha \,|\, \sigma)$. Let Z be an oriented 1-dimensional polyhedron (a line segment) of length $\mathrm{len}(\sigma)$, partitioned into segments $Z_1, \ldots, Z_m$, with $\mathrm{len}(Z_i) = \mathrm{len}(\sigma_i)$. Let $\sigma' : \mathbf{I}^1 \to Z$ be the unique oriented linear bijection. There is a unique piecewise linear map $f : Z \to X$, such that $\sigma_i = f \circ \sigma' \circ \sigma_i^k$ as piecewise linear maps $\mathbf{I}^1 \to X$ for every i. Let $\alpha' := f^*(\alpha) \in \Omega^1_{\mathrm{pws}}(Z) \otimes \mathfrak{g}$. According to Propositions 3.3.22(3), 3.5.7 and 3.5.9 we have

$$\mathrm{MI}(\alpha \,|\, \sigma) = \mathrm{MI}(\alpha' \,|\, \sigma').$$

And clearly

$$\int_\sigma \alpha = \int_{\sigma'} \alpha'.$$

Because the piecewise linear map $f : Z \to X$ is a linear metric embedding on each of its linear pieces, it follows that $\|\alpha'\|_{\mathrm{Sob}} \leq \|\alpha\|_{\mathrm{Sob}}$; and hence we can choose $\epsilon_1(\alpha') \geq \epsilon_1(\alpha)$ and $c_1(\alpha') \leq c_1(\alpha)$.

Note that

$$\mathrm{len}(\sigma') = \mathrm{len}(Z) = \mathrm{len}(\sigma) < \epsilon_1(\alpha) \,.$$

So we can replace X with Z, σ with σ' and α with α'. Doing so, we can now assume that $m = 1$ and σ is a single linear map.

Step 2. Here we assume that σ is an α-tiny linear map, and we let $\epsilon := \mathrm{len}(\sigma)$. Take any $k \geq 0$. We know that

$$\mathrm{RP}_k(\alpha \,|\, \sigma) = \prod_{i=1}^{2^k} \mathrm{RP}_0(\alpha \,|\, \sigma \circ \sigma_i^k).$$

Also (by Lemma 3.3.1) we have

$$\mathrm{RP}_k(\alpha \,|\, \sigma), \mathrm{RP}_0(\alpha \,|\, \sigma \circ \sigma_i^k) \in V_0(G).$$

For any i let

$$\lambda_i := \log_G (\mathrm{RP}_0(\alpha \,|\, \sigma \circ \sigma_i^k)) \in \mathfrak{g},$$

so

$$\mathrm{RP}_k(\alpha \,|\, \sigma) = \prod_{i=1}^{2^k} \exp_G(\lambda_i).$$

Let us write

$$\mathrm{RS}_k(\alpha \,|\, \sigma) := \sum_{i=1}^{2^k} \lambda_i.$$

This is a Riemann sum for the usual integral.

By Lemma 3.3.1 we have

$$\|\lambda_i\| \le c_1(\alpha) \cdot (\tfrac{1}{2})^k \cdot \epsilon$$

for every i. Using property (ii) of Theorem 2.1.2 we see that

$$\| \log_G(\mathrm{RP}_k(\alpha \,|\, \sigma)) - \mathrm{RS}_k(\alpha \,|\, \sigma) \|$$
$$\le c_0(G) \cdot \left(\sum_{i=1}^{2^k} \|\lambda_i\| \right)^2 \le c_0(G) \cdot c_1(\alpha)^2 \cdot \epsilon^2 .$$

In the limit $k \to \infty$ we have

$$\lim_{k \to \infty} \mathrm{RS}_k(\alpha \,|\, \sigma) = \int_\sigma \alpha$$

and

$$\lim_{k \to \infty} \mathrm{RP}_k(\alpha \,|\, \sigma) = \mathrm{MI}(\alpha \,|\, \sigma),$$

so the proof is done. □

Corollary 3.5.12. *Let $\alpha \in \Omega^1_{\mathrm{pws}}(X) \otimes \mathfrak{g}$, and let σ be an α-tiny closed string in X which bounds a polygon Z. Then*

$$\| \log_G(\mathrm{MI}(\alpha \,|\, \sigma)) \| \le c_0(G) \cdot c_1(\alpha)^2 \cdot \mathrm{len}(\sigma)^2 + \mathrm{area}(Z) \cdot \|\alpha\|_{\mathrm{Sob}} .$$

Proof. Choose an orientation on Z. By the abelian Stokes Theorem (Theorem 1.8.3) we have

$$\left\| \int_\sigma \alpha \right\| = \left\| \int_Z \mathrm{d}(\alpha) \right\| \le \mathrm{area}(Z) \cdot \|\alpha\|_{\mathrm{Sob}} .$$

Now combine this estimate with the Proposition above. □

Proposition 3.5.13. *Let $\Phi : G \to \mathrm{GL}_m(\mathbb{R})$ be a representation, and let $\alpha \in \Omega^1_{\mathrm{pws}}(X) \otimes \mathfrak{g}$. Then there are constants $\epsilon_5(\alpha, \Phi)$ and $c_5(\alpha, \Phi)$ such that*

$$0 < \epsilon_5(\alpha, \Phi) \le 1 \quad and \quad c_5(\alpha, \Phi) \ge 1,$$

and such that conditions (i)-(iii) below hold for every string σ in X satisfying $\mathrm{len}(\sigma) < \epsilon_5(\alpha, \Phi)$. Let us write

$$\alpha' := \mathrm{Lie}(\Phi)(\alpha) \in \Omega^1_{\mathrm{pws}}(X) \otimes \mathfrak{gl}_n(\mathbb{R}),$$

and

$$g' := \mathrm{MI}(\alpha' \mid \sigma) = \Phi\big(\mathrm{MI}(\alpha \mid \sigma)\big) \in \mathrm{GL}_m(\mathbb{R}).$$

Let **1** *be the identity operator on* $\mathbb{R}^m$, *and let* $\|{-}\|$ *denote the operator norm on* $\mathbb{R}^m$. *The conditions are:*

(i)

$$\|g' - \mathbf{1}\| \le c_5(\alpha, \Phi) \cdot \mathrm{len}(\sigma).$$

(ii)

$$\left\| g' - \left(1 + \int_\sigma \alpha'\right) \right\| \le c_5(\alpha, \Phi) \cdot \mathrm{len}(\sigma)^2.$$

(iii) *Assume* α' *is smooth. Let* x_0 *be the initial point of* σ, *and let* $\alpha'(x_0)$ *be the constant form defined in Definition 1.3.5. Then*

$$\left\| g' - \left(1 + \int_\sigma \alpha'(x_0)\right) \right\| \le c_5(\alpha, \Phi) \cdot \mathrm{len}(\sigma)^2.$$

Proof. Let us write $H := \mathrm{GL}_m(\mathbb{R})$ and $\mathfrak{h} := \mathfrak{gl}_n(\mathbb{R})$. Define $\epsilon_5(\alpha, \Phi) := \epsilon_1(\alpha')$, in the sense of Definition 3.3.16, and let $d := c_1(\alpha') \cdot \epsilon_1(\alpha')$.

By reasons of convergence of analytic functions on compact domains, there is a constant c such that for every matrix $\lambda \in \mathfrak{h}$ with $\|\lambda\| \le d$ the inequalities

(3.5.14)
$$\| \exp_H(\lambda) - \mathbf{1} \| \le c \cdot \|\lambda\|$$

and

(3.5.15)
$$\| \exp_H(\lambda) - (\mathbf{1} + \lambda) \| \le c \cdot \|\lambda\|^2$$

hold.

Take a string σ with $\epsilon := \mathrm{len}(\sigma) < \epsilon_5(\alpha, \Phi)$. By Proposition 3.5.8(3) we have $g' \in V_0(H)$ and

$$\|\lambda\| \le c_1(\alpha') \cdot \epsilon \le d,$$

for the elements $g' := \mathrm{MI}(\alpha' \mid \sigma)$ and $\lambda := \log_H(g') \in \mathfrak{h}$. Inequality (3.5.14) gives

(3.5.16)
$$\|g' - \mathbf{1}\| \le c \cdot c_1(\alpha') \cdot \epsilon.$$

Next, using Proposition 3.5.11 and inequality (3.5.15) we have

(3.5.17)
$$\left\| g' - \left(1 + \int_\sigma \alpha'\right) \right\| \le \left\| g' - (\mathbf{1} + \lambda) \right\| + \left\| \lambda - \int_\sigma \alpha' \right\|$$
$$\le c \cdot c_1(\alpha')^2 \cdot \epsilon^2 + c_1(\alpha')^2 \cdot \epsilon^2.$$

Finally, assume that α' is smooth. By Taylor expansion of the coefficients of α' (cf. (3.3.7)) we have the estimate

$$\| \alpha'(x) - \alpha'(x_0) \| \le \|\alpha'\|_{\mathrm{Sob}} \cdot \epsilon$$

for every point x in the image of the string σ. Therefore

$$
(3.5.18) \qquad \left\| \int_\sigma \alpha' - \int_\sigma \alpha'(x_0) \right\| \le \|\alpha'\|_{\text{Sob}} \cdot \epsilon^2 .
$$

From the inequalities (3.5.16), (3.5.17) and (3.5.18) we can now easily extract a constant $c_5(\alpha, \Phi)$ for which all three conditions hold. ☐

4 Multiplicative Integration in Dimension Two

We pass up to dimension 2. Here it turns out that things are really much more complicated, for geometrical reasons.

A rudimentary multiplicative integration on surfaces was already introduced by Schlesinger in 1928, in his nonabelian 2-dimensional Stokes Theorem (see [DF, Appendix A.II,9]). However we need stronger results, for which a more complicated multiplicative integration procedure is required.

It turns out (this was already in Schlesinger's work) that the correct multiplicative integral is twisted: there is a 2-form, say β, that is integrated, but this integration is twisted by a 1-form α. The geometric cycles on which integration is performed are the kites, to be defined now.

4.1 Kites

By a *pointed polyhedron* (X, x_0) we mean a polyhedron X (see Chapter 1), together with a base point $x_0 \in X$. As base point for $\mathbf{I}^2$ we always take the vertex v_0; cf. (1.2.1).

Definition 4.1.1. Let (X, x_0) be a pointed polyhedron.

(1) A *quadrangular kite* in (X, x_0) is a pair (σ, τ), where σ is a string in X (see Definition 3.5.1), and $\tau : \mathbf{I}^2 \to X$ is a linear map. The conditions are that $\sigma(v_0) = x_0$ and $\sigma(v_1) = \tau(v_0)$.

(2) If $\tau(\mathbf{I}^2)$ is 2-dimensional then we call (σ, τ) a *nondegenerate kite*.

(3) If $\tau(\mathbf{I}^2)$ is a square in X (of positive size), then we call (σ, τ) a *square kite*.

See Figure 8 for illustration.

Until Chapter 9, where triangular kites are introduced, we shall only encounter quadrangular kites. Hence in Chapters 4-8 a kite shall always mean a quadrangular kite.

Consider a kite (σ, τ). The image $\tau(\mathbf{I}^2)$ is a parallelogram in X. We denote the area of $\tau(\mathbf{I}^2)$ by area(τ). If $\tau(\mathbf{I}^2)$ is a square, then we denote the side of this square by side(τ).

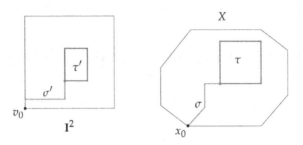

FIGURE 8. A linear quadrangular kite (σ, τ) in the pointed polyhedron (X, x_0), and a linear quadrangular kite (σ', τ') in the pointed polyhedron $(\mathbf{I}^2, v_0)$.

We view $(\mathbf{I}^2, v_0)$ as an oriented pointed polyhedron. Suppose (σ, τ) is a nondegenerate kite in $(\mathbf{I}^2, v_0)$. If the orientation of τ is positive, then we say that the kite (σ, τ) is positively oriented.

We shall need the following composition operation on kites. Suppose (σ_1, τ_1) is a kite in (X, x_0), and (σ_2, τ_2) is a kite in $(\mathbf{I}^2, v_0)$. Then $\tau_1 \circ \tau_2 : \mathbf{I}^2 \to X$ is a linear map, and $\sigma_1 * (\tau_1 \circ \sigma_2)$ is a string in X. (See (3.5.2) for the concatenation operation $*$.) We define

$$(4.1.2) \qquad (\sigma_1, \tau_1) \circ (\sigma_2, \tau_2) := \big(\sigma_1 * (\tau_1 \circ \sigma_2), \tau_1 \circ \tau_2\big),$$

which is also a kite in (X, x_0). Note that this composition operation is associative. For an illustration see Figures 8 and 9.

Let (σ, τ) be a kite in (X, x_0), and let $f : (X, x_0) \to (Y, y_0)$ be a piecewise linear map between pointed polyhedra. Assume that the restriction of f to the subpolyhedron $\tau(\mathbf{I}^2) \subset X$ is linear. As in formula (3.5.5) we have a string $f \circ \sigma$ in Y. We define

$$(4.1.3) \qquad f \circ (\sigma, \tau) := (f \circ \sigma, f \circ \tau),$$

which is a kite in (Y, y_0).

Given a kite (σ, τ) in (X, x_0), its *boundary* is the closed string $\partial(\sigma, \tau)$ defined as follows. First we define

$$(4.1.4) \qquad \partial \mathbf{I}^2 := (v_0, v_1) * (v_1, (1,1)) * ((1,1), v_2) * (v_2, v_0).$$

This is a closed string in $\mathbf{I}^2$, based at v_0. Next we let $\partial \tau := \tau \circ (\partial \mathbf{I}^2)$, where composition is in the sense of (3.5.5). So $\partial \tau$ is a closed string in X. Finally we define

$$(4.1.5) \qquad \partial(\sigma, \tau) := \sigma * (\partial \tau) * \sigma^{-1},$$

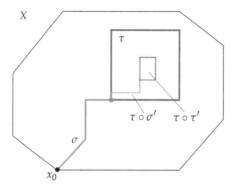

FIGURE 9. Continued from Figure 8: The linear quadrangular kite $(\sigma, \tau) \circ (\sigma', \tau')$ in the pointed polyhedron (X, x_0).

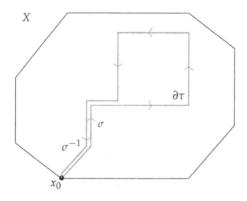

FIGURE 10. The boundary $\partial(\sigma, \tau) = \sigma * \partial \tau * \sigma^{-1}$ of the kite (σ, τ) from Figure 8.

where σ^{-1} is the inverse string from (3.5.4). See Figure 10.

Here is a useful fact about the geometry of kites.

Proposition 4.1.6. *Let (σ, τ) be a kite in (X, x_0). Then there is a square kite (σ', τ') in $(\mathbf{I}^2, v_0)$, and a piecewise linear map of pointed polyhedra $f : (\mathbf{I}^2, v_0) \to (X, x_0)$, such that $\mathrm{len}(\sigma') \leq 1$, f is linear on $\tau'(\mathbf{I}^2)$, and*

$$(\sigma, \tau) = f \circ (\sigma', \tau')$$

as kites in (X, x_0).

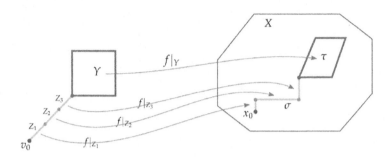

FIGURE 11. The map $f|_{Y\cup Z} : Y \cup Z \to X$, where $Z = Z_1 \cup Z_2 \cup Z_3$.

Proof. Say $\sigma = (\sigma_1, \ldots, \sigma_m)$ is the decomposition of σ into pieces. If $m = 0$ (i.e. σ is the empty string), then we let σ' also be the empty string, and we take $\tau' := \mathrm{id}_{I^2}$ and $f := \tau$.

Otherwise, if $m > 0$, then we define the square kite (σ', τ') in (I^2, v_0) as follows. The square map $\tau' : I^2 \to I^2$ is defined on vertices by the formula
$$\tau'(v_0, v_1, v_2) := \left(\left(\tfrac{1}{2}, \tfrac{1}{2}\right), \left(1, \tfrac{1}{2}\right), \left(\tfrac{1}{2}, 1\right)\right).$$
And we let $Y := \tau'(I^2)$.

Next consider the oriented line segment Z going from v_0 to $\left(\tfrac{1}{2}, \tfrac{1}{2}\right)$. We divide Z into m equal pieces, labeled $Z_1, \ldots, Z_m$. We let $\sigma'_i : I^1 \to Z$ be the positively oriented linear map with image Z_i. And we let σ' be the string $\sigma' := (\sigma'_1, \ldots, \sigma'_m)$.

The map $f|_Y$ is defined to be the unique linear map $Y \to X$ such that $f \circ \tau' = \tau$. And for every i the map $f|_{Z_i}$ defined to be the unique piecewise linear map $Z_i \to X$ such that $f \circ \sigma'_i = \sigma_i$. We thus have a map $f|_{Y\cup Z} : Y \cup Z \to X$; see Figure 11 for an illustration.

Finally let $g : I^2 \to Y \cup Z$ be any piecewise linear retraction; for instance as suggested by Figure 12. We define $f := f|_{Y\cup Z} \circ g : I^2 \to X$. See Figure 13. $\square$

4.2 Binary Tessellations of I^2

For $k \geq 0$ the *k-th binary subdivision of* I^2 is the cellular subdivision $\mathrm{sd}^k I^2$ of I^2 into 4^k squares, each of side $\left(\tfrac{1}{2}\right)^k$. The 1-skeleton of $\mathrm{sd}^k I^2$ is the set $\mathrm{sk}_1 \mathrm{sd}^k I^2$ consisting of the union of all edges (i.e. 1-cells) in $\mathrm{sd}^k I^2$. Thus $\mathrm{sk}_1 \mathrm{sd}^k I^2$ is a grid. See Figure 14.

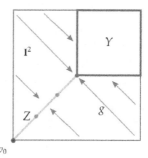

FIGURE 12. A piecewise linear retraction $g : \mathbf{I}^2 \to Y \cup Z$.

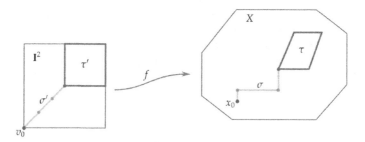

FIGURE 13. The piecewise linear map $f := f|_{Y \cup Z} \circ g$, and the kite (σ', τ') such that $f \circ (\sigma', \tau') = (\sigma, \tau)$.

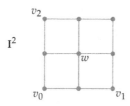

FIGURE 14. The 1-st binary subdivision $\mathrm{sd}^1 \mathbf{I}^2$ of $\mathbf{I}^2$. Here $w := \left(\frac{1}{2}, \frac{1}{2}\right)$.

Definition 4.2.1. Let k be a natural number.

   (1) Let $p = 0, 1, 2$. A linear map $\sigma : I^p \to I^2$ is said to be *patterned on* $\mathrm{sd}^k I^2$ if the image $\sigma(I^p)$ is a p-cell of $\mathrm{sd}^k I^2$.

   (2) A string $\sigma = (\sigma_1, \ldots, \sigma_m)$ in I^2 is said to be patterned on $\mathrm{sd}^k I^2$ if each piece $\sigma_i : I^1 \to I^2$ is a linear map patterned on $\mathrm{sd}^k I^2$.

   (3) A square kite (σ, τ) in (I^2, v_0) is said to be patterned on $\mathrm{sd}^k I^2$ if both the linear map τ and the string σ are patterned on $\mathrm{sd}^k I^2$.

The fundamental group of the topological space $\mathrm{sk}_1 \mathrm{sd}^k I^2$, based at v_0, is denoted by $\pi_1(\mathrm{sk}_1 \mathrm{sd}^k I^2)$. It is a free group on 4^k generators. Given a closed string σ patterned on $\mathrm{sd}^k I^2$ and based at v_0, we denote by $[\sigma]$ the corresponding element of $\pi_1(\mathrm{sk}_1 \mathrm{sd}^k I^2)$. In particular, if (σ, τ) is a kite patterned on $\mathrm{sd}^k I^2$, then the boundary $\partial(\sigma, \tau)$ represents an element

$$[\partial(\sigma, \tau)] \in \pi_1(\mathrm{sk}_1 \mathrm{sd}^k I^2).$$

Recall the boundary ∂I^2 from equation (4.1.4).

Definition 4.2.2. Let k be a natural number. A *tessellation of* I^2 *patterned on* $\mathrm{sd}^k I^2$ is a sequence

$$\rho = ((\sigma_1, \tau_1), \ldots, (\sigma_{4^k}, \tau_{4^k}))$$

of kites in (I^2, v_0), satisfying these two conditions:

   (i) Each kite (σ_i, τ_i) is patterned on $\mathrm{sd}^k I^2$.

   (ii) One has

$$\prod_{i=1}^{4^k} [\partial(\sigma_i, \tau_i)] = [\partial I^2]$$

   in the group $\pi_1(\mathrm{sk}_1 \mathrm{sd}^k I^2)$. The product is according to the convention (2.1.1).

Remark 4.2.3. Suppose ρ is a tessellation of I^2 patterned on $\mathrm{sd}^k I^2$. Then, in the notation of the definition, each kite (σ_i, τ_i) is positively oriented, and each 2-cell of $\mathrm{sd}^k I^2$ occurs as $\tau_i(I^2)$ for exactly one index i.

   This assertion (that we will not use in the book) can be proved directly, by a topological argument. But it also follows from Corollary 7.5.2, by taking the abelian Lie groups $G := 1$ and $H := \mathrm{GL}_1(\mathbb{R})$, and the differential forms $\alpha := 0$ and $\beta := f \cdot dt_1 \wedge dt_1$, where $f : I^2 \to \mathbb{R}$ is a smooth nonnegative bump function supported in the interior of a given 2-cell of $\mathrm{sd}^k I^2$.

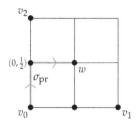

FIGURE 15. The probe σ_{pr} in $\mathbf{I}^2$. Here $w := (\frac{1}{2}, \frac{1}{2})$.

The *probe* is the string

$$\sigma_{\mathrm{pr}} := \left(v_0, \left(0, \tfrac{1}{2}\right)\right) * \left(\left(0, \tfrac{1}{2}\right), \left(\tfrac{1}{2}, \tfrac{1}{2}\right)\right) \tag{4.2.4}$$

(with two pieces) in $\mathbf{I}^2$. See Figure 15 for an illustration and Remark 4.3.4 for an explanation.

Definition 4.2.5. Let k be a natural number. The *k-th binary tessellation of* $\mathbf{I}^2$ is the sequence

$$\mathrm{tes}^k \mathbf{I}^2 = \left(\mathrm{tes}_1^k \mathbf{I}^2, \ldots, \mathrm{tes}_{4^k}^k \mathbf{I}^2\right) = \left((\sigma_1^k, \tau_1^k), \ldots, (\sigma_{4^k}^k, \tau_{4^k}^k)\right)$$

of kites in $(\mathbf{I}^2, v_0)$, patterned on $\mathrm{sd}^k \mathbf{I}^2$, that is defined recursively as follows.

(1) For $k = 0$ let σ_1^0 be the empty string, and let τ_1^0 be the identity map of $\mathbf{I}^2$.

(2) For $k = 1$ all four strings σ_i^1 are the same; they are $\sigma_i^1 := \sigma_{\mathrm{pr}}$. The four linear maps $\tau_i^1 : \mathbf{I}^2 \to \mathbf{I}^2$ are patterned on $\mathrm{sd}^1 \mathbf{I}^2$, positively oriented, and have $\tau_i^1(v_0) = (\frac{1}{2}, \frac{1}{2})$. It remains to specify the points $\tau_i^1(v_1)$:

$$\tau_1^1(v_1) = (\tfrac{1}{2}, 0), \; \tau_2^1(v_1) = (1, \tfrac{1}{2}), \; \tau_3^1(v_1) = (\tfrac{1}{2}, 1), \; \tau_4^1(v_1) = (0, \tfrac{1}{2}).$$

(3) For $k \geq 2$ we define

$$\mathrm{tes}^k \mathbf{I}^2 := (\mathrm{tes}^1 \mathbf{I}^2) \circ (\mathrm{tes}^{k-1} \mathbf{I}^2).$$

Here composition of sequences of kites is using the operations (4.1.2) and (3.1.1).

See Figures 16 and 17 for an illustration.

It is clear that sequence $\mathrm{tes}^k \mathbf{I}^2$ is a tessellation of $\mathbf{I}^2$ patterned on $\mathrm{sd}^k \mathbf{I}^2$, in the sense of Definition 4.2.2. We call (σ_1^0, τ_1^0) the *basic kite*.

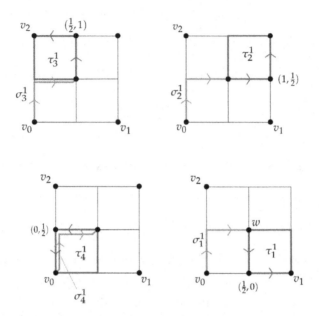

FIGURE 16. The 1-st binary tessellation $\mathrm{tes}^1\,\mathbf{I}^2$. The arrow-heads indicate the orientation of the linear maps.

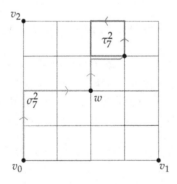

FIGURE 17. The kite (σ_7^2, τ_7^2) in $\mathrm{tes}^2\,\mathbf{I}^2$.

Hopefully there will be no confusion between the linear map σ_i^k belonging $\mathrm{tes}^k\,\mathbf{I}^1$, and the string σ_i^k belonging $\mathrm{tes}^k\,\mathbf{I}^2$; these are distinct objects that share the same notation.

An easy calculation shows that

(4.2.6) $$\mathrm{len}(\sigma_i^k) \leq 2$$

for all k and i.

Definition 4.2.7. Let (σ, τ) be a kite in (X, x_0). For $k \in \mathbb{N}$ and $i \in \{1, \ldots, 4^k\}$ let

$$\mathrm{tes}_i^k(\sigma, \tau) := (\sigma, \tau) \circ \mathrm{tes}_i^k \mathbf{I}^2 = (\sigma, \tau) \circ (\sigma_i^k, \tau_i^k),$$

which is also a kite in (X, x_0). The sequence of kites

$$\mathrm{tes}^k(\sigma, \tau) := \left(\mathrm{tes}_1^k(\sigma, \tau), \ldots, \mathrm{tes}_{4^k}^k(\sigma, \tau) \right)$$

is called the k-*th binary tessellation* of (σ, τ).

Remark 4.2.8. The choice of strings for the kites in the binary tessellations (and the ordering of the kites) is clearly artificial, and also very asymmetrical. As we shall see later, in favorable situations this will not matter at all – any other tessellation works! See Corollary 7.5.2.

4.3   Additive Twisting and Riemann Products

Let $\mathfrak{h}$ be a finite dimensional vector space (over $\mathbb{R}$). We denote by $\mathrm{GL}(\mathfrak{h})$ the group of linear automorphisms of $\mathfrak{h}$, which is a Lie group (noncanonically isomorphic to $\mathrm{GL}_d(\mathbb{R})$ for $d := \dim \mathfrak{h}$).

Definition 4.3.1. A *twisting setup* is the data $\mathbf{C} = (G, H, \Psi_{\mathfrak{h}})$, consisting of:

(1) Lie groups G and H, with Lie algebras $\mathfrak{g}$ and $\mathfrak{h}$ respectively.
(2) A map of Lie groups $\Psi_{\mathfrak{h}} : G \to \mathrm{GL}(\mathfrak{h})$, called an *additive twisting*.

Warning: we do not assume that the map $\Psi_{\mathfrak{h}}(g) : \mathfrak{h} \to \mathfrak{h}$, for $g \in G$, is an automorphism of Lie algebras!

Example 4.3.2. Let G be any Lie group. Take $H := G$ and $\Psi_{\mathfrak{h}} := \mathrm{Ad}_{\mathfrak{h}}$, the adjoint action of $H = G$ on its Lie algebra. Then $(G, H, \Psi_{\mathfrak{h}})$ is a twisting setup. Here $\Psi_{\mathfrak{h}}(g)$ is in fact an automorphism of Lie groups.

Let us fix, for the rest of this chapter, a twisting setup $\mathbf{C} = (G, H, \Psi_{\mathfrak{h}})$. The Lie algebras of G and H are $\mathfrak{g}$ and $\mathfrak{h}$ respectively.

We choose some euclidean norm $\|-\|_{\mathfrak{g}}$ on the vector space $\mathfrak{g}$. As in Chapter 2 we also choose an open neighborhood $V_0(G)$ of 1 in G on which $\log_G$ is well-defined, a convergence radius $\epsilon_0(G)$, and a commutativity constant $c_0(G)$. Likewise we choose $\|-\|_{\mathfrak{h}}$, $V_0(H)$, $\epsilon_0(H)$ and $c_0(H)$. Given $g \in G$, the linear operator $\Psi_{\mathfrak{h}}(g) \in \mathrm{End}(\mathfrak{h})$ is given the operator norm $\|\Psi_\mathfrak{h}(g)\|$. It should be noted that these choices are auxiliary only, and do not effect the definition of the multiplicative integration.

Piecewise smooth differential forms were discussed in Section 1.6. The string σ_{pr} (the probe) was introduced in (4.2.4).

Definition 4.3.3 (Basic Riemann Product). Let (X, x_0) be a pointed polyhedron, let

$$\alpha \in \Omega^1_{\mathrm{pws}}(X) \otimes \mathfrak{g},$$

let

$$\beta \in \Omega^2_{\mathrm{pws}}(X) \otimes \mathfrak{h},$$

and let (σ, τ) be a kite in (X, x_0). We define an element

$$\mathrm{RP}_0(\alpha, \beta \,|\, \sigma, \tau) \in H,$$

called the *basic Riemann product of* (α, β) *on* (σ, τ), as follows. Write $Z := \tau(\mathbf{I}^2)$ and $z := \tau(\frac{1}{2}, \frac{1}{2}) \in Z$. There are two cases to consider:

(1) Assume $\dim Z = 2$ and z is a smooth point of the form $\beta|_Z$. Put on Z the orientation compatible with τ. Choose a triangle Y in Z such that $z \in \mathrm{Int}\, Y$ and $\beta|_Y$ is smooth, and let $\tilde\beta \in \mathcal{O}(Y) \otimes \mathfrak{h}$ be the coefficient of $\beta|_Y$ with respect to the orientation form of Y (see Definition 1.8.2). Also let

$$g := \mathrm{MI}(\alpha \,|\, \sigma * (\tau \circ \sigma_{\mathrm{pr}})) \in G.$$

We define

$$\mathrm{RP}_0(\alpha, \beta \,|\, \sigma, \tau) := \exp_H\big(\mathrm{area}(Z) \cdot \Psi_\mathfrak{h}(g)(\tilde\beta(z))\big).$$

(2) If $\dim Z < 2$, or $\dim Z = 2$ and z is a singular point of $\beta|_Z$, we define

$$\mathrm{RP}_0(\alpha, \beta \,|\, \sigma, \tau) := 1.$$

It is obvious that case (1) of the definition is independent of the triangle Y. See Figure 18 for an illustration.

Remark 4.3.4. We call the string σ_{pr} "the probe" because it reaches into the middle of the square $\mathbf{I}^2$. The "reading" it gives, namely the formula for $\mathrm{RP}_0(\alpha, \beta \,|\, \sigma, \tau)$ in part (1) of the definition above, is better than (0.8.1), because it converges to the limit faster: order of $\mathrm{side}(\tau)^4$ versus $\mathrm{side}(\tau)^3$. Cf. Lemma 4.4.4 below.

Suppose we are given a finite sequence of kites

$$\rho = \big((\sigma_1, \tau_1), \ldots, (\sigma_m, \tau_m)\big)$$

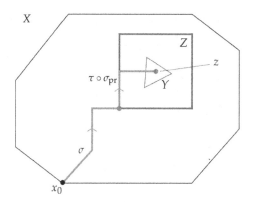

FIGURE 18. Calculating $\mathrm{RP}_0(\alpha, \beta \mid \sigma, \tau)$ in the smooth case. Here $Z = \tau(\mathbf{I}^2)$, $z = \tau(\frac{1}{2}, \frac{1}{2})$, and Y is a triangle in Z such that $\beta|_Y$ is smooth.

in (X, x_0). We write

$$(4.3.5) \qquad \mathrm{RP}_0(\alpha, \beta \mid \rho) := \prod_{i=1}^{m} \mathrm{RP}_0(\alpha, \beta \mid \sigma_i, \tau_i).$$

Recall the binary tessellation $\mathrm{tes}^k(\sigma, \tau)$ from Definition 4.2.7.

Definition 4.3.6 (Refined Riemann Product). Let (X, x_0) be a pointed polyhedron, let $\alpha \in \Omega^1_{\mathrm{pws}}(X) \otimes \mathfrak{g}$, let $\beta \in \Omega^2_{\mathrm{pws}}(X) \otimes \mathfrak{h}$, and let (σ, τ) be a kite in (X, x_0). For $k \geq 0$ we define

$$\mathrm{RP}_k(\alpha, \beta \mid \sigma, \tau) := \mathrm{RP}_0(\alpha, \beta \mid \mathrm{tes}^k(\sigma, \tau)) = \prod_{i=1}^{4^k} \mathrm{RP}_0(\alpha, \beta \mid (\sigma, \tau) \circ (\sigma_i^k, \tau_i^k)).$$

Lemma 4.3.7. *Let* $f : (X', x_0') \to (X, x_0)$ *be a piecewise linear map of pointed polyhedra, and let* (σ', τ') *be a kite in* (X', x_0'). *Assume that* f *is linear on* $\tau'(\mathbf{I}^2)$, *and let*

$$(\sigma, \tau) := f \circ (\sigma', \tau'),$$

which is a kite in (X, x_0). *Let* $\alpha' := f^*(\alpha)$ *and* $\beta' := f^*(\beta)$. *Then*

$$\mathrm{RP}_k(\alpha, \beta \mid \sigma, \tau) = \mathrm{RP}_k(\alpha', \beta' \mid \sigma', \tau')$$

for any $k \geq 0$.

Proof. Take $k \geq 0$ and $i \in \{1, \ldots, 4^k\}$. We will prove that

$$(4.3.8) \qquad \mathrm{RP}_0(\alpha, \beta \mid (\sigma, \tau) \circ (\sigma_i^k, \tau_i^k)) = \mathrm{RP}_0(\alpha', \beta' \mid (\sigma', \tau') \circ (\sigma_i^k, \tau_i^k)).$$

Let $Z := \tau(\mathbf{I}^2) \subset X$ and $Z' := \tau'(\mathbf{I}^2) \subset X'$. If $\dim Z < 2$ then $\beta|_Z = 0$ and $\beta'|_{Z'} = 0$, and hence both sides of (4.3.8) equal 1.

If $\dim Z = 2$ then the linear map $f|_{Z'} : Z' \to Z$ is bijective. Let $w := (\tfrac{1}{2}, \tfrac{1}{2}) \in \mathbf{I}^2$, $z_i := (\tau \circ \tau_i^k)(w) \in Z$ and $z_i' := (\tau' \circ \tau_i^k)(w) \in Z'$. Then z_i is a smooth point of $\beta|_Z$ if and only if z_i' is a smooth point of $\beta'|_{Z'}$. In the singular case again both sides of (4.3.8) equal 1.

In the smooth case we know (by Proposition 3.5.9) that

$$\mathrm{MI}(\alpha \,|\, \sigma * (\tau \circ \sigma_i^k) * (\tau \circ \tau_i^k \circ \sigma_{\mathrm{pr}})) = \mathrm{MI}(\alpha' \,|\, \sigma' * (\tau' \circ \sigma_i^k) * (\tau' \circ \tau_i^k \circ \sigma_{\mathrm{pr}})).$$

This says the twistings are the same. Let $\tilde{\beta}_i$ be the coefficient of β near z_i, as in case (1) of Definition 4.3.3; and let $\tilde{\beta}_i'$ be the coefficient of β' near z_i'. Then

$$\mathrm{area}(Z_i) \cdot \tilde{\beta}_i(z_i) = \mathrm{area}(Z_i') \cdot \tilde{\beta}_i'(z_i').$$

So in this case we also get equality in (4.3.8). $\qquad\square$

4.4 Convergence of Riemann Products

We continue with the setup of Section 4.3. Fix differential forms $\alpha \in \Omega^1_{\mathrm{pws}}(X) \otimes \mathfrak{g}$ and $\beta \in \Omega^2_{\mathrm{pws}}(X) \otimes \mathfrak{h}$.

Lemma 4.4.1. *There are constants $c_1(\alpha, \beta)$ and $\epsilon_1(\alpha, \beta)$ with the following properties.*

    (i) *The inequalities below hold:*

$$1 \leq c_1(\alpha, \beta)$$
$$0 < \epsilon_1(\alpha, \beta) \leq 1$$
$$\epsilon_1(\alpha, \beta) \cdot c_1(\alpha, \beta) \leq \tfrac{1}{4}\epsilon_0(H).$$

    (ii) *Suppose (σ, τ) is a square kite in (X, x_0) such that $\mathrm{side}(\tau) < \epsilon_1(\alpha, \beta)$ and $\mathrm{len}(\sigma) \leq 4 \cdot \mathrm{diam}(X)$. Then for any $k \geq 0$ one has*

$$\mathrm{RP}_k(\alpha, \beta \,|\, \sigma, \tau) \in V_0(H)$$

    and

$$\| \log_H(\mathrm{RP}_k(\alpha, \beta \,|\, \sigma, \tau)) \| \leq c_1(\alpha, \beta) \cdot \mathrm{side}(\tau)^2.$$

Proof. Let $w := (\tfrac{1}{2}, \tfrac{1}{2}) \in \mathbf{I}^2$. Given a square kite (σ, τ), let $Z := \tau(\mathbf{I}^2)$, $z := \tau(w)$ and $\epsilon := \mathrm{side}(\tau)$.

For $i \in \{1, \ldots, 4^k\}$ define

(4.4.2) $$(\sigma_i, \tau_i) := (\sigma, \tau) \circ (\sigma_i^k, \tau_i^k).$$

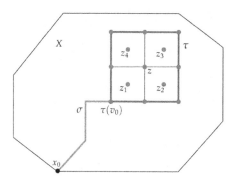

FIGURE 19. The kite (σ, τ) in (X, x_0) and the points $z_1, \ldots, z_{4^k}$ in $\tau(\mathbf{I}^2)$. Here $k = 1$.

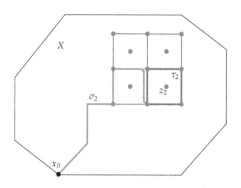

FIGURE 20. The kite (σ_2, τ_2) in (X, x_0).

This is a square kite in (X, x_0) satisfying $\mathrm{side}(\tau_i) = (\frac{1}{2})^k \epsilon$ and $\mathrm{area}(\tau_i) = (\frac{1}{4})^k \epsilon^2$. Let $Z_i := \tau_i(\mathbf{I}^2)$ and $z_i := \tau_i(w)$. See Figures 19 and 20 for illustration. Since $\mathrm{len}(\sigma_i^k) \leq 2$ and $\epsilon \leq \mathrm{diam}(X)$, we have

(4.4.3) $\mathrm{len}(\sigma_i * (\tau_i \circ \sigma_{\mathrm{pr}})) = \mathrm{len}(\sigma) + \epsilon \cdot \mathrm{len}(\sigma_i^k) + (\frac{1}{2})^k \epsilon \leq 7 \cdot \mathrm{diam}(X).$

Consider the group elements

$$g_i := \mathrm{MI}(\alpha \,|\, \sigma_i * (\tau_i \circ \sigma_{\mathrm{pr}})) \in G$$

and

$$h_i := \mathrm{RP}_0(\alpha, \beta \,|\, \sigma_i, \tau_i) \in H.$$

By definition of the Riemann product we have

$$RP_k(\alpha, \beta \mid \sigma, \tau) = \prod_{i=1}^{4^k} h_i.$$

Take $i \in \{1, \ldots, 4^k\}$. If z_i is a smooth point of $\beta|_Z$ (this is the good case), then let $\tilde{\beta}_i$ be the coefficient of $\beta|_Z$ near z_i, and let

$$\lambda_i := (\tfrac{1}{4})^k \cdot \epsilon^2 \cdot \Psi_{\mathfrak{h}}(g_i)(\tilde{\beta}_i(z_i)) \in \mathfrak{h}.$$

Otherwise, if z_i is a singular point of $\beta|_Z$ (this is the bad case), then we let $\lambda_i := 0$. In any case, by definition we have $h_i = \exp_H(\lambda_i)$.

According to Proposition 3.4.3 and the inequality (4.4.3) we have

$$\|\Psi_{\mathfrak{h}}(g_i)\| \leq \exp\left(c_4(\alpha, \Psi_{\mathfrak{h}}) \cdot 7 \cdot \mathrm{diam}(X)\right)$$

for some constant $c_4(\alpha, \Psi_{\mathfrak{h}})$. Note that this bound is independent of k and (σ, τ).

Let

$$c := \exp\left(c_4(\alpha, \Psi_{\mathfrak{h}}) \cdot 7 \cdot \mathrm{diam}(X)\right) \cdot \|\beta\|_{\mathrm{Sob}} + 1,$$

$$\epsilon_1(\alpha, \beta) := \epsilon_0(H)^{\frac{1}{2}} \cdot c^{-\frac{1}{2}}$$

and

$$c_1(\alpha, \beta) := c \cdot (c_0(H) + 1).$$

Now assume our kite satisfies $\epsilon < \epsilon_1(\alpha, \beta)$. Because $\|\tilde{\beta}_i(z_i)\| \leq \|\beta\|_{\mathrm{Sob}}$ in the good case, and $\lambda_i = 0$ in the bad case, we see that

$$\|\lambda_i\| \leq (\tfrac{1}{4})^k \cdot \epsilon^2 \cdot c < (\tfrac{1}{4})^k \cdot \epsilon_0(H).$$

Hence $\sum_{i=1}^{4^k} \|\lambda_i\| < \epsilon_0(H)$, and by property (ii) of Theorem 2.1.2 we can deduce that

$$\prod_{i=1}^{4^k} h_i \in V_0(H)$$

and

$$\left\| \log_H \left(\prod_{i=1}^{4^k} h_i \right) \right\| \leq c_0(H) \cdot \left(\sum_{i=1}^{4^k} \|\lambda_i\| \right) \leq c_0(H) \cdot \epsilon^2 \cdot c \leq c_1(\alpha, \beta) \cdot \epsilon^2.$$

$\square$

Lemma 4.4.4. *There are constant $\epsilon_2(\alpha, \beta)$ and $c_2(\alpha, \beta)$ with the following properties:*

(i) *$c_2(\alpha, \beta) \geq c_1(\alpha, \beta)$ and $0 < \epsilon_2(\alpha, \beta) \leq \epsilon_1(\alpha, \beta)$.*

(ii) *Suppose (σ, τ) is a square kite in (X, x_0) such that $\beta|_{\tau(\mathbf{I}^2)}$ is smooth, $\mathrm{side}(\tau) < \epsilon_2(\alpha, \beta)$ and $\mathrm{len}(\sigma) \leq 4 \cdot \mathrm{diam}(X)$. Then*

$$\left\| \log_H \left(RP_k(\alpha, \beta \mid \sigma, \tau) \right) - \log_H \left(RP_0(\alpha, \beta \mid \sigma, \tau) \right) \right\| \leq c_2(\alpha, \beta) \cdot \mathrm{side}(\tau)^4$$

for every $k \geq 0$.

FIGURE 21. A string π in $Z = \tau(I^2)$, with initial point z_0 and terminal point z.

The exponent 4 in "side$(\tau)^4$" in the inequality above will be of utmost importance later on. On the other hand, the factor 4 appearing in "$4 \cdot \text{diam}(X)$" is quite arbitrary (any number bigger than 2 would probably do just as well).

Proof. Let $w := (\frac{1}{2}, \frac{1}{2}) \in I^2$. Suppose (σ, τ) is some square kite in (X, x_0). Write $Z := \tau(I^2)$, $z_0 := \tau(w) \in Z$ and $\epsilon := \text{side}(Z)$. Assume that $\epsilon > 0$ and $\beta|_Z$ is smooth (otherwise there is nothing to prove). Put on Z the orientation compatible with τ.

Let $\tilde{\beta} \in \mathcal{O}(Z) \otimes \mathfrak{h}$ be the coefficient of $\beta|_Z$, as in Definition 1.8.2. Let

$$g_0 := \text{MI}(\alpha \,|\, \sigma * (\tau \circ \sigma_{\text{pr}})) \in G,$$

and

$$\lambda_0 := \epsilon^2 \cdot \Psi_{\mathfrak{h}}(g_0)(\tilde{\beta}(z_0)) \in \mathfrak{h}.$$

So by definition

$$\exp_H(\lambda_0) = \text{RP}_0(\alpha, \beta \,|\, \sigma, \tau).$$

Suppose π is some string in Z, with initial point z_0. Let z be the terminal point of π. See Figure 21.

We shall need the following variant of the Taylor expansion of $\tilde{\beta}(z)$ to second order around z_0:

$$\tilde{\beta}(z) = \tilde{\beta}(z_0) + \int_{\pi} (d\tilde{\beta})(z_0) + R^2(\tilde{\beta}, \pi).$$

Here $d\tilde{\beta} \in \Omega^1(Z) \otimes \mathfrak{h}$, and $(d\tilde{\beta})(z_0) \in \Omega^1_{\text{const}}(Z) \otimes \mathfrak{h}$ is the associated constant form (see Definition 1.3.5). Thus $\int_{\pi} (d\tilde{\beta})(z_0)$ is the linear term in the expansion – it depends linearly on π. And the quadratic remainder term $R^2(\tilde{\beta}, \pi) \in \mathfrak{h}$ has this bound:

$$\|R^2(\tilde{\beta}, \pi)\| \leq \|\beta\|_{\text{Sob}} \cdot \text{len}(\pi)^2.$$

Therefore we get the estimate

(4.4.5) $$\| \tilde{\beta}(z) - (\tilde{\beta}(z_0) + \int_{\pi} (d\tilde{\beta})(z_0)) \| \leq \|\beta\|_{\text{Sob}} \cdot \text{len}(\pi)^2 .$$

Next let

$$g := \text{MI}(\alpha \mid \pi) \in G$$

and

$$\alpha' := \text{Lie}(\Psi_{\mathfrak{h}})(\alpha) \in \Omega^1_{\text{pws}}(X) \otimes \text{End}(\mathfrak{h}).$$

Assume that $\text{len}(\pi) < \epsilon_5(\alpha, \Psi_{\mathfrak{h}})$. According to Proposition 3.5.13 we have this estimate for the operator $\Psi_{\mathfrak{h}}(g) \in \text{End}(\mathfrak{h})$:

(4.4.6) $$\| \Psi_{\mathfrak{h}}(g) - (1 + \int_{\pi} \alpha'(z_0)) \| \leq c_5(\alpha, \Psi_{\mathfrak{h}}) \cdot \text{len}(\pi)^2 .$$

And by Proposition 3.5.10 we have the bound

(4.4.7) $$\| \Psi_{\mathfrak{h}}(g) \| \leq \exp(c_4(\alpha, \Psi_{\mathfrak{h}}) \cdot \epsilon_5(\alpha, \Psi_{\mathfrak{h}})) .$$

By combining inequalities (4.4.5), (4.4.6) and (4.4.7), we see that there exists a constant $c(\alpha, \beta)$ such that

(4.4.8) $$\| \Psi_{\mathfrak{h}}(g)(\tilde{\beta}(z)) - \left(\tilde{\beta}(z_0) + (\int_{\pi} \alpha'(z_0))(\tilde{\beta}(z_0)) + \int_{\pi} (d\tilde{\beta})(z_0) \right) \|$$
$$\leq c(\alpha, \beta) \cdot \text{len}(\pi)^2 .$$

This holds for every string π in Z with $\text{len}(\pi) < \epsilon_5(\alpha, \Psi_{\mathfrak{h}})$, $\pi(v_0) = z_0$ and $\pi(v_1) = z$.

Now take $k \geq 0$. For any index $i \in \{1, \ldots, 4^k\}$ let π_i be the unique string in Z such that

$$(\tau \circ \sigma_{\text{pr}}) * \pi_i = (\tau \circ \sigma_i^k) * (\tau \circ \tau_i^k \circ \sigma_{\text{pr}})$$

as strings. This is a string with initial point z_0, and with terminal point

$$z_i := (\tau \circ \tau_i^k)(w).$$

Notice that for $i \leq 2^k$ the strings π_i and π_{2^k+i} are reflections of each other relative to the point z_0. See Figure 22.

Define

$$g_i := \text{MI}(\alpha \mid \pi_i) \in G$$

and

$$\lambda_i := (\tfrac{1}{4})^k \cdot \epsilon^2 \cdot \Psi_{\mathfrak{h}}(g_0 \cdot g_i)(\tilde{\beta}(z_i)) \in \mathfrak{h}.$$

So we have

$$\exp_H(\lambda_i) = \text{RP}_0(\alpha, \beta \mid (\sigma, \tau) \circ (\sigma_i^k, \tau_i^k)).$$

And the bound for λ_i is

(4.4.9) $$\|\lambda_i\| \leq (\tfrac{1}{4})^k \cdot \epsilon^2 \cdot c'(\alpha, \beta) ,$$

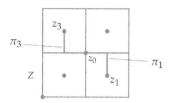

FIGURE 22. The strings π_1 and π_3 in Z. Here $k = 1$.

where we write

$$c'(\alpha, \beta) := \exp(c_4(\alpha, \Psi_\mathfrak{h}) \cdot 7 \cdot \operatorname{diam}(X)) \cdot \|\beta\|_{\mathrm{Sob}} + 1.$$

Using the abbreviation

$$\gamma := \epsilon^2 \cdot \Psi_\mathfrak{h}(g_0)(\mathrm{d}\tilde{\beta}) \in \Omega^1_{\mathrm{pws}}(Z) \otimes \mathfrak{h},$$

formula (4.4.8), and the inequality $\operatorname{len}(\pi_i) \leq \epsilon$, we obtain

(4.4.10)
$$\left\| \lambda_i - (\tfrac{1}{4})^k \cdot \left(\lambda_0 + \left(\int_{\pi_i} \alpha'(z_0) \right)(\lambda_0) + \int_{\pi_i} \gamma(z_0) \right) \right\|$$
$$\leq c(\alpha, \beta) \cdot (\tfrac{1}{4})^k \cdot \epsilon^4.$$

For this to be true we should assume that $\epsilon < \epsilon_5(\alpha, \Psi_\mathfrak{h})$.

Let us set

$$\epsilon_2(\alpha, \beta) := \min\big(\epsilon_1(\alpha, \beta), \, c'(\alpha, \beta)^{-1/2} \cdot \epsilon_0(H)^{1/2}, \, \epsilon_5(\alpha, \Psi_\mathfrak{h})\big),$$

We now assume furthermore that $\epsilon < \epsilon_2(\alpha, \beta)$. In particular, from (4.4.9) we obtain

$$\sum_{i=1}^{4^k} \|\lambda_i\| \leq \epsilon_0(H).$$

We know that

$$\prod_{i=1}^{4^k} \exp_H(\lambda_0) = \mathrm{RP}_k(\alpha, \beta \mid \sigma, \tau).$$

Therefore we can use property (ii) of Theorem 2.1.2 to deduce that

$$\left\| \log_H\big(\mathrm{RP}_k(\alpha, \beta \mid \sigma, \tau)\big) - \sum_{i=1}^{4^k} \lambda_i \right\| \leq c_0(H) \cdot \epsilon^4 \cdot c'(\alpha, \beta)^2.$$

The geometric symmetry of the sequence of strings $\pi_1, \ldots, \pi_{4^k}$ implies that

$$\left(\int_{\pi_i} \alpha'(z_0) \right)(\lambda_0) = -\left(\int_{\pi_{2^k+i}} \alpha'(z_0) \right)(\lambda_0)$$

for $i \leq 2^k$; and therefore

$$\sum_{i=1}^{4^k} \left(\int_{\pi_i} \alpha'(z_0) \right)(\lambda_0) = 0.$$

Similarly

$$\sum_{i=1}^{4^k} \int_{\pi_i} \gamma(z_0) = 0.$$

Plugging in the estimate (4.4.10) we obtain

$$\left\| \sum_{i=1}^{4^k} \lambda_i - \lambda_0 \right\| \leq c(\alpha, \beta) \cdot \epsilon^4.$$

We see that the constant

$$c_2(\alpha, \beta) := c(\alpha, \beta) + c_0(H) \cdot c'(\alpha, \beta)^2 + c_1(\alpha, \beta)$$

works.   □

Definition 4.4.11. Let us fix constants $\epsilon_2(\alpha, \beta)$ and $c_2(\alpha, \beta)$ as in Lemma 4.4.4. A square kite (σ, τ) in (X, x_0) will be called (α, β)-*tiny* in this chapter if

$$\mathrm{side}(\tau) < \epsilon_2(\alpha, \beta)$$

and

$$\mathrm{len}(\sigma) \leq 4 \cdot \mathrm{diam}(X).$$

Definition 4.4.12. Let (σ, τ) be a nondegenerate kite in (X, x_0). For $k \in \mathbb{N}$ and $i \in \{1, \ldots, 4^k\}$ let $Z_i := (\tau \circ \tau_i^k)(\mathbf{I}^2) \subset X$. An index i is called *good* if the forms $\alpha|_{Z_i}$ and $\beta|_{Z_i}$ are smooth. Otherwise i is called *bad*. The sets of good and bad indices are denoted by $\mathrm{good}(\tau, k)$ and $\mathrm{bad}(\tau, k)$ respectively.

Lemma 4.4.13. *Let Z be a 2-dimensional subpolyhedron of X. There exist constants $a_0(\alpha, \beta, Z)$ and $a_1(\alpha, \beta, Z)$ such that for any nondegenerate kite (σ, τ) in (X, x_0) with $\tau(\mathbf{I}^2) \subset Z$, one has*

$$|\mathrm{bad}(\tau, k)| \leq a_0(\alpha, \beta, Z) + a_1(\alpha, \beta, Z) \cdot 2^k.$$

Proof. Let $Z_1, \ldots, Z_m$ be line segments in Z such that the singular locus of α and the singular locus of β are contained in $\bigcup_{j=1}^m Z_j$. Take $a_0(\alpha, \beta, Z) := 2m$ and $a_1(\alpha, \beta, Z) := 2 \cdot \sum_{j=1}^m \mathrm{len}(Z_j)$.   □

Lemma 4.4.14. *Let Z be a 2-dimensional subpolyhedron of X. Then there is a constant $c_3(\alpha, \beta, Z) \geq 0$, such that for any (α, β)-tiny kite (σ, τ) satisfying $\tau(\mathbf{I}^2) \subset Z$, and any $k' \geq k \geq 0$, one has*

$$\| \log_H(\mathrm{RP}_{k'}(\alpha, \beta \,|\, \sigma, \tau)) - \log_H(\mathrm{RP}_k(\alpha, \beta \,|\, \sigma, \tau)) \|$$
$$\leq (\tfrac{1}{4})^k \cdot c_3(\alpha, \beta, Z) \cdot \mathrm{side}(\tau)^2.$$

Proof. Let (σ, τ) an (α, β)-tiny kite such that $\tau(\mathbf{I}^2) \subset Z$. Write $\epsilon := \mathrm{side}(\tau)$. For $i \in \{1, \dots, 4^k\}$ let (σ_i, τ_i) be as in equation (4.4.2), and let $Z_i := \tau_i(\mathbf{I}^2) \subset Z$. Note that $\mathrm{side}(Z_i) = (\tfrac{1}{2})^k \cdot \epsilon$.

Let $l := k' - k$. If $i \in \mathrm{good}(\tau, k)$, then by Lemma 4.4.4 we know that

$$\| \log_H(\mathrm{RP}_l(\alpha, \beta \,|\, \sigma_i, \tau_i)) - \log_H(\mathrm{RP}_0(\alpha, \beta \,|\, \sigma_i, \tau_i)) \| \leq c_2(\alpha, \beta) \cdot ((\tfrac{1}{2})^k \cdot \epsilon)^4.$$

If $i \in \mathrm{bad}(\tau, k)$, then by Lemma 4.4.1 we know that

$$\| \log_H(\mathrm{RP}_l(\alpha, \beta \,|\, \sigma_i, \tau_i)) - \log_H(\mathrm{RP}_0(\alpha, \beta \,|\, \sigma_i, \tau_i)) \| \leq 2c_1(\alpha, \beta) \cdot ((\tfrac{1}{2})^k \cdot \epsilon)^2.$$

Therefore by property (iv) of Theorem 2.1.2 and Lemma 4.4.13 we have

$$\| \log_H(\mathrm{RP}_{k'}(\alpha, \beta \,|\, \sigma, \tau)) - \log_H(\mathrm{RP}_k(\alpha, \beta \,|\, \sigma, \tau)) \|$$
$$= \| \log_H(\textstyle\prod_{i=1}^{4^k} \mathrm{RP}_l(\alpha, \beta \,|\, \sigma_i, \tau_i) - \log_H(\textstyle\prod_{i=1}^{4^k} \mathrm{RP}_0(\alpha, \beta \,|\, \sigma_i, \tau_i) \|$$
$$\leq c_0(H) \cdot \big(|\mathrm{good}(\tau, k)| \cdot c_2(\alpha, \beta) \cdot (\tfrac{1}{2})^{4k} \cdot \epsilon^4$$
$$+ |\mathrm{bad}(\tau, k)| \cdot 2c_1(\alpha, \beta) \cdot (\tfrac{1}{2})^{2k} \cdot \epsilon^2 \big)$$
$$\leq c_3(\alpha, \beta, Z) \cdot (\tfrac{1}{2})^k \cdot \epsilon^2,$$

where we take (very generously)

$$c_3(\alpha, \beta, Z) := c_0(H) \cdot \big(c_2(\alpha, \beta) + 2(a_0(\alpha, \beta, Z) + a_1(\alpha, \beta, Z)) \cdot c_1(\alpha, \beta) \big).$$

$\square$

Theorem 4.4.15. *Let (X, x_0) be a pointed polyhedron, let $\alpha \in \Omega^1_{\mathrm{pws}}(X) \otimes \mathfrak{g}$, let $\beta \in \Omega^2_{\mathrm{pws}}(X) \otimes \mathfrak{h}$, and let (σ, τ) be a kite in (X, x_0). Then the limit*

$$\lim_{k \to \infty} \mathrm{RP}_k(\alpha, \beta \,|\, \sigma, \tau)$$

exists in H.

Proof. According to Proposition 4.1.6 and Lemma 4.3.7 we can assume that $(X, x_0) = (\mathbf{I}^2, v_0)$ and $\mathrm{len}(\sigma) \leq 1$.

Take k large enough such that for each $i \in \{1, \dots, 4^k\}$ the kite (σ_i, τ_i), in the notation of (4.4.2), is (α, β)-tiny. For any $k' \geq 0$ we have

$$\mathrm{RP}_{k+k'}(\alpha, \beta \,|\, \sigma, \tau) = \prod_{i=1}^{4^k} \mathrm{RP}_{k'}(\alpha, \beta \,|\, \sigma_i, \tau_i).$$

Thus it suffices to prove that for any i the limit

$$\lim_{k'\to\infty} \mathrm{RP}_{k'}(\alpha,\beta \,|\, \sigma_i, \tau_i)$$

exists. Now Lemma 4.4.14 says that the sequence

$$\left(\mathrm{RP}_{k'}(\alpha,\beta \,|\, \sigma_i, \tau_i)\right)_{k'\geq 0}$$

is a Cauchy sequence in H; and therefore it converges.   □

Definition 4.4.16 (Multiplicative Integral). Let $(G, H, \Psi_\mathfrak{h})$ be a twisting setup, let (X, x_0) be a pointed polyhedron, let $\alpha \in \Omega^1_{\mathrm{pws}}(X) \otimes \mathfrak{g}$, let $\beta \in \Omega^2_{\mathrm{pws}}(X) \otimes \mathfrak{h}$, and let (σ, τ) be a kite in (X, x_0). We define the *multiplicative integral of β twisted by α on (σ, τ)* to be

$$\mathrm{MI}(\alpha,\beta \,|\, \sigma, \tau) := \lim_{k\to\infty} \mathrm{RP}_k(\alpha,\beta \,|\, \sigma, \tau) \in H.$$

If $(X, x_0) = (\mathbf{I}^2, v_0)$ then we write

$$\mathrm{MI}(\alpha,\beta \,|\, \mathbf{I}^2) := \mathrm{MI}(\alpha,\beta \,|\, \sigma^0_1, \tau^0_1),$$

where (σ^0_1, τ^0_1) is the basic kite.

4.5   Some Properties of MI

We continue with the setup of the previous sections.

Proposition 4.5.1. *In the situation of Definition 4.4.16, for any $k \geq 0$ one has*

$$\mathrm{MI}(\alpha,\beta \,|\, \sigma, \tau) = \prod_{i=1}^{4^k} \mathrm{MI}(\alpha,\beta \,|\, (\sigma, \tau) \circ (\sigma^k_i, \tau^k_i)).$$

Proof. For any $k' \geq 0$ we have

$$\mathrm{RP}_{k+k'}(\alpha,\beta \,|\, \sigma, \tau) = \prod_{i=1}^{4^k} \mathrm{RP}_{k'}(\alpha,\beta \,|\, (\sigma, \tau) \circ (\sigma^k_i, \tau^k_i)).$$

Now take the limit $\lim_{k'\to\infty}$.   □

Proposition 4.5.2. *Consider the situation of Definition 4.4.16, and assume that (σ, τ) is an (α, β)-tiny kite.*

(1) *One has*

$$\mathrm{MI}(\alpha,\beta \,|\, \sigma, \tau) \in V_0(H),$$

and

$$\left\| \log_H\left(\mathrm{MI}(\alpha,\beta \,|\, \sigma, \tau)\right) \right\| \leq c_1(\alpha,\beta) \cdot \mathrm{side}(\tau)^2.$$

(2) If $\beta|_{\tau(\mathbf{I}^2)}$ is smooth then

$$\| \log_H(\mathrm{MI}(\alpha, \beta \,|\, \sigma, \tau)) - \log_H(\mathrm{RP}_0(\alpha, \beta \,|\, \sigma, \tau)) \| \leq c_2(\alpha, \beta) \cdot \mathrm{side}(\tau)^4.$$

(3) Let Z be a 2-dimensional subpolyhedron of X containing $\tau(\mathbf{I}^2)$. For any $k \geq 0$ one has

$$\| \log_H(\mathrm{MI}(\alpha, \beta \,|\, \sigma, \tau)) - \log_H(\mathrm{RP}_k(\alpha, \beta \,|\, \sigma, \tau)) \|$$
$$\leq (\tfrac{1}{2})^k \cdot c_3(\alpha, \beta, Z) \cdot \mathrm{side}(\tau)^2.$$

Proof. Immediate from Lemmas 4.4.1, 4.4.4 and 4.4.14.   □

Proposition 4.5.3 (Functoriality in X). *Let $f : (Y, y_0) \to (X, x_0)$ be a piecewise linear map between pointed polyhedra, let $\alpha \in \Omega^1_{\mathrm{pws}}(X) \otimes \mathfrak{g}$, let $\beta \in \Omega^2_{\mathrm{pws}}(X) \otimes \mathfrak{h}$, and let (σ, τ) be a kite in (Y, y_0). Assume that f is linear on $\tau(\mathbf{I}^2)$. Then*

$$\mathrm{MI}(\alpha, \beta \,|\, f \circ \sigma, f \circ \tau) = \mathrm{MI}(f^*(\alpha), f^*(\beta) \,|\, \sigma, \tau).$$

Proof. Immediate from Lemma 4.3.7.   □

The next proposition says that "in the tiny scale the 2-dimensional MI is abelian".

Proposition 4.5.4. *There are constants $\epsilon_{2'}(\alpha, \beta)$ and $c_{2'}(\alpha, \beta)$ with the following properties:*

(i) *$c_{2'}(\alpha, \beta) \geq c_2(\alpha, \beta)$ and $0 < \epsilon_{2'}(\alpha, \beta) \leq \epsilon_2(\alpha, \beta)$.*

(ii) *Suppose (σ, τ) is a square kite in (X, x_0) such that $\mathrm{side}(\tau) < \epsilon_{2'}(\alpha, \beta)$ and $\mathrm{len}(\sigma) \leq 4 \cdot \mathrm{diam}(X)$. Let $g := \mathrm{MI}(\alpha \,|\, \sigma) \in G$. Then*

$$\| \log_H(\mathrm{MI}(\alpha, \beta \,|\, \sigma, \tau)) - \Psi_{\mathfrak{h}}(g)(\int_\tau \beta) \| \leq c_{2'}(\alpha, \beta) \cdot \mathrm{side}(\tau)^3.$$

Actually with more effort we can get a better estimate (order $\mathrm{side}(\tau)^4$) in property (ii) above.

Proof. This is very similar to Proposition 3.5.11. Take

$$\epsilon_{2'}(\alpha, \beta) := \min(\epsilon_2(\alpha, \beta), \epsilon_5(\alpha, \Psi_{\mathfrak{h}})),$$

where $\epsilon_5(\alpha, \Psi_{\mathfrak{h}})$ is the constant from Proposition 3.5.13. Let us write $\epsilon := \mathrm{side}(\tau)$ and $Z := \tau(\mathbf{I}^2)$. Assume that $\epsilon < \epsilon_{2'}(\alpha, \beta)$.

Take $k \geq 0$. For $i \in \{1, \dots, 4^k\}$ let

$$(\sigma_i, \tau_i) := (\sigma, \tau) \circ (\sigma_i^k, \tau_i^k) = \mathrm{tes}_i^k(\sigma, \tau).$$

Write $Z_i := \tau_i(\mathbf{I}^2)$ and $z_i := \tau_i(w)$, where $w := (\tfrac{1}{2}, \tfrac{1}{2}) \in \mathbf{I}^2$ as usual. The sets $\mathrm{good}(\tau, k)$ and $\mathrm{bad}(\tau, k)$ were defined in Definition 4.4.12.

Define

$$g_0' := \mathrm{MI}(\alpha \,|\, \tau \circ \sigma_{\mathrm{pr}}) \in G,$$

$$g_i' := \mathrm{MI}(\alpha \,|\, (\tau \circ \sigma_i^k) * (\tau \circ \tau_i^k \circ \sigma_{\mathrm{pr}}))$$

and

$$\lambda_i := \log_H (\mathrm{RP}_0(\alpha, \beta \,|\, \sigma_i, \tau_i)) \in \mathfrak{h}$$

for $i \in \{1, \ldots, 4^k\}$. From Definition 4.3.3 and Proposition 3.5.10 we know that

$$(4.5.5) \qquad \|\lambda_i\| \le (\tfrac{1}{4})^k \cdot \epsilon^2 \cdot \|\Psi_\mathfrak{h}(g \cdot g_i')\| \cdot \|\beta\|_{\mathrm{Sob}} \le (\tfrac{1}{4})^k \cdot \epsilon^2 \cdot c,$$

where we let

$$c := \exp\big(c_4(\alpha, \phi) \cdot 6 \operatorname{diam}(X)\big) \cdot \|\beta\|_{\mathrm{Sob}}.$$

According to property (ii) of Theorem 2.1.2 we have

$$(4.5.6) \qquad \begin{aligned} &\Big\| \log_H (\mathrm{RP}_k(\alpha, \beta \,|\, \sigma, \tau)) - \sum_{i=1}^{4^k} \lambda_i \Big\| \\ &\le c_0(H) \cdot \Big(\sum_{i=1}^{4^k} \|\lambda_i\| \Big)^2 \le c_0(H) \cdot (\tfrac{1}{4})^{2k} \cdot \epsilon^4 \cdot c^2. \end{aligned}$$

For $i \in \mathrm{good}(\tau, k)$ let $\tilde{\beta}_i$ be the coefficient of $\beta|_{z_i}$. Define

$$\mu_i := \begin{cases} (\tfrac{1}{4})^k \cdot \epsilon^2 \cdot \Psi_\mathfrak{h}(g)(\tilde{\beta}_i(z_i)) & \text{if } i \text{ is good}, \\ 0 & \text{otherwise} \end{cases}$$

and

$$\mathrm{RS}_k(\alpha, \beta \,|\, \sigma, \tau) := \sum_{i=1}^{4^k} \mu_i.$$

Now let us compare μ_i to λ_i. If i is a bad index, then

$$\|\mu_i - \lambda_i\| = \|\lambda_i\| \le (\tfrac{1}{4})^k \cdot \epsilon^2 \cdot c$$

by (4.5.5). On the other hand, if i is a good index then

$$\mu_i = \Psi_\mathfrak{h}(g \cdot g_0' \cdot g_i'^{-1} \cdot g^{-1})(\lambda_i).$$

By Proposition 3.5.10 the operators $\Psi_\mathfrak{h}(g)$ and $\Psi_\mathfrak{h}(g^{-1})$ have known bounds (since the length of the string σ is bounded by $4 \cdot \operatorname{diam}(X)$). Hence there is a bound for the conjugation operator $\mathrm{Ad}(\Psi_\mathfrak{h}(g))$ on $\mathrm{End}(\mathfrak{h})$. And by Proposition 3.5.13 there is an estimate for the norm of the operator $\Psi_\mathfrak{h}(g_0' \cdot g_i'^{-1}) - \mathbf{1}$, where $\mathbf{1}$ denotes the identity operator of $\mathfrak{h}$. Since $\mathrm{Ad}(\Psi_\mathfrak{h}(g))$ fixes $\mathbf{1}$, we can conclude that there is a constant c' (independent of (σ, τ) or k) such that

$$\begin{aligned} &\big\| \Psi_\mathfrak{h}(g \cdot g_0' \cdot g_i'^{-1} \cdot g^{-1}) - \mathbf{1} \big\| \\ &= \big\| \mathrm{Ad}(\Psi_\mathfrak{h}(g))\big(\Psi_\mathfrak{h}(g_0' \cdot g_i'^{-1}) - \mathbf{1}\big) \big\| \le c' \cdot \epsilon. \end{aligned}$$

Hence using (4.5.5) we get

$$\| \mu_i - \lambda_i \| \leq c' \cdot c \cdot (\tfrac{1}{4})^k \cdot \epsilon^3 .$$

Summing over all i we see that

$$\left\| \sum_{i=1}^{4^k} \lambda_i - \sum_{i=1}^{4^k} \mu_i \right\| \leq \sum_{i=1}^{4^k} \| \mu_i - \lambda_i \|$$

$$\leq |\mathrm{good}(\tau,k)| \cdot c' \cdot c \cdot (\tfrac{1}{4})^k \cdot \epsilon^3 + |\mathrm{bad}(\tau,k)| \cdot (\tfrac{1}{4})^k \cdot \epsilon^2 \cdot c$$

$$\leq 4^k \cdot c' \cdot c \cdot (\tfrac{1}{4})^k \cdot \epsilon^3 + (a_0 + a_1 \cdot 2^k) \cdot (\tfrac{1}{4})^k \cdot \epsilon^2 \cdot c .$$

Here $a_0 := a_0(\alpha, \beta, Z)$ and $a_1 := a_1(\alpha, \beta, Z)$ are the constants from Lemma 4.4.13. Combining this estimate with (4.5.6) we obtain

(4.5.7)
$$\| \log_H (\mathrm{RP}_k(\alpha, \beta \,|\, \sigma, \tau)) - \mathrm{RS}_k(\alpha, \beta \,|\, \sigma, \tau) \|$$
$$\leq c_0(H) \cdot (\tfrac{1}{4})^{2k} \cdot \epsilon^4 \cdot c^2 + 4^k \cdot c' \cdot c \cdot (\tfrac{1}{4})^k \cdot \epsilon^3$$
$$+ (a_0 + a_1 \cdot 2^k) \cdot (\tfrac{1}{4})^k \cdot \epsilon^2 \cdot c .$$

Finally, by properties of the usual Riemann integration we have

$$\lim_{k \to \infty} \mathrm{RS}_k(\alpha, \beta \,|\, \sigma, \tau) = \Psi_{\mathfrak{h}}(g) \Big(\int_\tau \beta \Big).$$

Hence in the limit $k \to \infty$ we get

$$\left\| \log_H (\mathrm{MI}(\alpha, \beta \,|\, \sigma, \tau)) - \Psi_{\mathfrak{h}}(g) \Big(\int_\tau \beta \Big) \right\| \leq c' \cdot c \cdot \epsilon^3 ,$$

and we can take $c_{2'}(\alpha, \beta) := c' \cdot c$.   □

4.6   Transfer of Twisting Setups

Suppose H' is another Lie group, with Lie algebra $\mathfrak{h}'$. The vector space of $\mathbb{R}$-linear maps $\mathfrak{h} \to \mathfrak{h}'$ is denoted by $\mathrm{Hom}(\mathfrak{h}, \mathfrak{h}')$. Consider the $\mathcal{O}_{\mathrm{pws}}(X)$-module $\mathcal{O}_{\mathrm{pws}}(X) \otimes \mathrm{Hom}(\mathfrak{h}, \mathfrak{h}')$. An element

$$\phi \in \mathcal{O}_{\mathrm{pws}}(X) \otimes \mathrm{Hom}_{\mathbb{R}}(\mathfrak{h}, \mathfrak{h}')$$

is called a *piecewise smooth family of linear maps* from $\mathfrak{h}$ to $\mathfrak{h}'$. Indeed, we may view ϕ as a piecewise smooth map

$$\phi : X \to \mathrm{Hom}(\mathfrak{h}, \mathfrak{h}').$$

For any point $x \in X$ there is a linear map $\phi(x) : \mathfrak{h} \to \mathfrak{h}'$.

Definition 4.6.1. Suppose $\mathbf{C} = (G, H, \Psi_{\mathfrak{h}})$ and $\mathbf{C}' = (G', H', \Psi_{\mathfrak{h}'}')$ are two twisting setups. A *transfer of twisting setups* from $\mathbf{C}$ to $\mathbf{C}'$, parametrized by (X, x_0), is the data

$$\Theta_X = (\Theta_G, \Theta_H, \Theta_{\mathfrak{h},X}),$$

consisting of:

(1) Maps of Lie groups $\Theta_G : G \to G'$ and $\Theta_H : H \to H'$.
(2) An element $\Theta_{\mathfrak{h},X} \in \mathcal{O}_{\mathrm{pws}}(X) \otimes \mathrm{Hom}(\mathfrak{h}, \mathfrak{h}')$.

The following condition is required:

(∗) The equality

$$\mathrm{Lie}(\Theta_H) = \Theta_{\mathfrak{h},X}(x_0)$$

holds in $\mathrm{Hom}(\mathfrak{h}, \mathfrak{h}')$.

We denote this transfer by $\Theta_X : \mathbf{C} \to \mathbf{C}'$.

Note that for $x \neq x_0$ the linear map $\phi(x) : \mathfrak{h} \to \mathfrak{h}'$ might fail to be a Lie algebra homomorphism.

Definition 4.6.2. Let $\Theta_X : \mathbf{C} \to \mathbf{C}'$ be a transfer of twisting setups as in Definition 4.6.1, and let $\alpha \in \Omega_{\mathrm{pws}}^1(X) \otimes \mathfrak{g}$. We say that α is a *connection compatible with* Θ_X if the following condition, called the *holonomy condition*, holds.

(◇) Let σ be a string in X, with $x_0 = \sigma(v_0)$ and $x_1 := \sigma(v_1)$. Define

$$g := \mathrm{MI}(\alpha \,|\, \sigma) \in G$$

and $g' := \Theta_G(g) \in G'$. Then the diagram

$$
\begin{array}{ccc}
\mathfrak{h} & \xleftarrow{\ \Psi_{\mathfrak{h}}(g)\ } & \mathfrak{h} \\
{\scriptstyle \Theta_{\mathfrak{h},X}(x_0)}\Big\downarrow & & \Big\downarrow{\scriptstyle \Theta_{\mathfrak{h},X}(x_1)} \\
\mathfrak{h}' & \xleftarrow{\ \Psi_{\mathfrak{h}'}'(g')\ } & \mathfrak{h}'
\end{array}
$$

is commutative.

Remark 4.6.3. The holonomy condition for α can be stated as a differential equation. We shall not need this equation in our treatment.

Consider a transfer of twisting setups Θ_X as above. The family of linear maps $\Theta_{\mathfrak{h},X}$ induces, by tensoring, a homomorphism of graded $\Omega_{\mathrm{pws}}(X)$-modules

(4.6.4) $\Theta_{\mathfrak{h},X} : \Omega_{\mathrm{pws}}(X) \otimes \mathfrak{h} \to \Omega_{\mathrm{pws}}(X) \otimes \mathfrak{h}'$.

Warning: usually $\Theta_{\mathfrak{h},X}$ does not commute with the de Rham operator d.

Proposition 4.6.5 (Functoriality in C). *Let $\Theta_X : \mathbf{C} \to \mathbf{C}'$ be a transfer of twisting setups, let $\alpha \in \Omega^1_{\mathrm{pws}}(X) \otimes \mathfrak{g}$, and let $\beta \in \Omega^2_{\mathrm{pws}}(X) \otimes \mathfrak{h}$. Assume that α is a connection compatible with Θ_X. We write*

$$\alpha' := \mathrm{Lie}(\Theta_G)(\alpha) \in \Omega^1_{\mathrm{pws}}(X) \otimes \mathfrak{g}'$$

and

$$\beta' := \Theta_{\mathfrak{h},X}(\beta) \in \Omega^2_{\mathrm{pws}}(X) \otimes \mathfrak{h}'.$$

Then for every kite (σ, τ) in (X, x_0) one has

$$\Theta_H\big(\mathrm{MI}(\alpha, \beta \,|\, \sigma, \tau)\big) = \mathrm{MI}(\alpha', \beta' \,|\, \sigma, \tau)$$

in H'.

Proof. Consider a kite (σ', τ') in $(\mathbf{I}^2, v_0)$ and a piecewise linear map f like in Proposition 4.1.6. By Proposition 4.5.3 we have

$$\mathrm{MI}(\alpha, \beta \,|\, \sigma, \tau) = \mathrm{MI}\big(f^*(\alpha), f^*(\beta) \,|\, \sigma', \tau'\big).$$

On the other hand $\big(\Theta_G, \Theta_H, f^*(\Theta_{\mathfrak{h},X})\big)$ is a transfer of twisting setups parametrized by $(\mathbf{I}^2, v_0)$, and by Proposition 4.5.3 we have

$$\mathrm{MI}\big(\alpha', \beta' \,|\, \sigma, \tau\big) = \mathrm{MI}\big(f^*(\alpha'), f^*(\beta') \,|\, \sigma', \tau'\big)$$
$$= \mathrm{MI}\big(\mathrm{Lie}(\Theta^1)(f^*(\alpha)), f^*(\Theta_{\mathfrak{h},X})(f^*(\beta)) \,|\, \sigma', \tau'\big).$$

Therefore we can assume that $(X, x_0) = (\mathbf{I}^2, v_0)$ and $\mathrm{len}(\sigma) \leq 1$. Using Proposition 4.5.1 we can further assume that (σ, τ) is (α, β)-tiny and (α', β')-tiny.

Fix $k \geq 0$. For $i \in \{1, \ldots, 4^k\}$ let

$$(\sigma_i, \tau_i) := (\sigma, \tau) \circ (\sigma_i^k, \tau_i^k).$$

Define

$$g_i := \mathrm{MI}(\alpha \,|\, \sigma_i * (\tau_i \circ \sigma_{\mathrm{pr}})) \in G$$

and

$$g_i' := \mathrm{MI}(\alpha' \,|\, \sigma_i * (\tau_i \circ \sigma_{\mathrm{pr}})) \in G'.$$

By Proposition 3.5.9 we have $g_i' = \Theta_G(g_i)$. Also define $Z_i := \tau_i(\mathbf{I}^2)$ and $z_i := \tau_i(\tfrac{1}{2}, \tfrac{1}{2})$.

Suppose that $i \in \mathrm{good}(\tau, k)$, with notation as in Definition 4.4.12. Then there is a function $\tilde{\beta}_i \in \mathcal{O}(Z_i) \otimes \mathfrak{h}$, called the coefficient of $\beta|_{Z_i}$, satisfying

$$\beta|_{Z_i} = \tilde{\beta}_i \cdot \mathrm{d}t_1 \wedge \mathrm{d}t_2.$$

The function

$$\tilde{\beta}_i' := \Theta_{\mathfrak{h},X}(\tilde{\beta}_i) \in \mathcal{O}(Z_i) \otimes \mathfrak{h}',$$

is then the coefficient of $\beta'|_{z_i}$. Note that

$$\tilde{\beta}'_i(z_i) = \Theta_{\mathfrak{h},X}(z_i)(\tilde{\beta}(z_i))$$

in $\mathfrak{h}'$. By condition $(*)$ of Definition 4.6.1 and condition $(\Diamond)$ of Definition 4.6.2 we have

$$\begin{aligned}
\Psi'_{\mathfrak{h}'}(g'_i)(\tilde{\beta}'_i(z_i)) &= \Psi'_{\mathfrak{h}'}(g'_i)(\Theta_{\mathfrak{h},X}(z_i)(\tilde{\beta}_i(z_i))) \\
&= \Theta_{\mathfrak{h},X}(x_0)(\Psi_{\mathfrak{h}}(g_i)(\tilde{\beta}_i(z_i))) \\
&= \mathrm{Lie}(\Theta_H)(\Psi_{\mathfrak{h}}(g_i)(\tilde{\beta}_i(z_i))).
\end{aligned}$$

By definition we have

$$\mathrm{RP}_0(\alpha,\beta \mid \sigma_i,\tau_i) = \exp_H((\tfrac{1}{4})^k \cdot \mathrm{area}(\tau) \cdot \Psi_{\mathfrak{h}}(g_i)(\tilde{\beta}_i(z_i)))$$

and

$$\mathrm{RP}_0(\alpha',\beta' \mid \sigma_i,\tau_i) = \exp_{H'}((\tfrac{1}{4})^k \cdot \mathrm{area}(\tau) \cdot \Psi_{\mathfrak{h}}(g'_i)(\tilde{\beta}'_i(z_i))).$$

Since

$$\exp_{H'} \circ \mathrm{Lie}(\Theta_H) = \Theta_H \circ \exp_H$$

we conclude that

(4.6.6) $$\mathrm{RP}_0(\alpha',\beta' \mid \sigma_i,\tau_i) = \Theta_H(\mathrm{RP}_0(\alpha,\beta \mid \sigma_i,\tau_i)).$$

Like in Lemma 4.4.13 we can find a bound for $|\mathrm{bad}(\tau,k)|$, and like in the proof of Lemma 4.4.14 we can estimate

$$\| \log_{H'}(\Theta_H(\mathrm{RP}_0(\alpha,\beta \mid \sigma_i,\tau_i))) - \log_{H'}(\mathrm{RP}_0(\alpha',\beta' \mid \sigma_i,\tau_i)) \|$$

when $i \in \mathrm{bad}(\tau,k)$. From these estimates and from (4.6.6) we conclude that there is a constant c, independent of k, such that

$$\| \log_{H'}(\Theta_H(\mathrm{RP}_k(\alpha,\beta \mid \sigma,\tau))) - \log_{H'}(\mathrm{RP}_k(\alpha',\beta' \mid \sigma,\tau)) \| \le c \cdot (\tfrac{1}{2})^k.$$

In the limit $k \to \infty$ we see that

$$\Theta_H(\mathrm{MI}(\alpha,\beta \mid \sigma,\tau)) = \mathrm{MI}(\alpha',\beta' \mid \sigma,\tau).$$

$$\square$$

5 Quasi Crossed Modules and Additive Feedback

The full strength of multiplicative integration requires a more elaborate setup than the twisting setup of Definition 4.3.1.

5.1 Quasi Crossed Modules

Let (Y, y_0) be a pointed analytic manifold. By *automorphism of pointed analytic manifolds* we mean an analytic diffeomorphism $f : Y \to Y$ such that $f(y_0) = y_0$. We denote by $\mathrm{Aut}(Y, y_0)$ the group of all such automorphisms.

Let G be a Lie group. An analytic action of G on (Y, y_0) by automorphisms of pointed manifolds is an analytic map $\Psi : G \times Y \to Y$ having the following properties. First, for any $g \in G$ the map $\Psi(g) : Y \to Y$, $\Psi(g)(y) := \Psi(g, y)$, is an automorphism of pointed analytic manifolds. Second, the function $\Psi : G \to \mathrm{Aut}(Y, y_0)$, $g \mapsto \Psi(g)$, is a group homomorphism.

Given an analytic action of G on (Y, y_0), and an element $g \in G$, the differential

$$\mathrm{d}_{y_0}(\Psi(g)) : T_{y_0} Y \to T_{y_0} Y$$

is an $\mathbb{R}$-linear automorphism of the tangent space $T_{y_0} Y$. In this way we get a map of Lie groups $G \to \mathrm{GL}(T_{y_0} Y)$, which we call the *linear action induced by* Ψ.

Let H be a Lie group, with unit element 1. We view it as a pointed analytic manifold $(H, 1)$.

Definition 5.1.1. A *Lie quasi crossed module* is the data

$$\mathbf{C} = (G, H, \Psi, \Phi_0)$$

consisting of:

  (1) Lie groups G and H.
  (2) An analytic action Ψ of G on H by automorphisms of pointed manifolds, called the *multiplicative twisting*.
  (3) A map of Lie groups $\Phi_0 : H \to G$, called the *multiplicative feedback*.
The condition is:

(∗) Consider Ψ as a group homomorphism $\Psi : G \to \mathrm{Aut}(H,1)$. Then there is equality

$$\Psi \circ \Phi_0 = \mathrm{Ad}_H$$

as group homomorphisms $H \to \mathrm{Aut}(H,1)$.

Remark 5.1.2. Let (G, H, Ψ, Φ_0) be a Lie quasi crossed module. Suppose G_0 is a closed Lie subgroup of G such that the following hold: $\Phi_0(H) \subset G_0$; $\Psi(g)$ is a group automorphism of H for any $g \in G_0$; and $\Phi_0 : H \to G_0$ is G_0-equivariant (relative to Ψ and Ad_{G_0}). Then (G_0, H, Ψ, Φ_0) is called a *Lie crossed module*. See [BM, BS]. In this situation condition (∗) is called the *Pfeiffer condition* in the literature.

Note that we can always find such a subgroup G_0: just take G_0 to be the closure of $\Phi_0(H)$ in G. An easy calculation shows that this subgroup has the required properties.

Here are three of examples of Lie quasi crossed modules.

Example 5.1.3. Suppose

$$1 \to N \to H \xrightarrow{\Phi_0} G \to 1$$

is a central extension of Lie groups. Since $\mathrm{Ad}_H(h)$ is trivial for $h \in N$, the action Ad_H induces an action of G on H, which we denote by Ψ. We get a Lie crossed module (G, H, Ψ, Φ_0).

Example 5.1.4. A very special case of Example 5.1.3 is when $H = G$ and $\Phi_0 = \mathrm{id}_G$. Namely

$$(G, H, \Psi, \Phi_0) = (G, G, \mathrm{Ad}_G, \mathrm{id}_G).$$

This is the situation dealt with in the classical work of Schlesinger.

Example 5.1.5. Let H be a *unipotent* Lie group, namely H is nilpotent and simply connected, and let $\mathfrak{h} := \mathrm{Lie}(H)$. The map $\exp_H : \mathfrak{h} \to H$ is then an analytic diffeomorphism. Take $G := \mathrm{GL}(\mathfrak{h})$. The canonical action of G on $\mathfrak{h}$ becomes, via $\exp_H$, and action of G on H by automorphisms of pointed manifolds, which we denote by Ψ. The adjoint action $\mathrm{Ad}_{\mathfrak{h}}$ of H on $\mathfrak{h}$ is a map of Lie groups $\Phi_0 : H \to G$. Then (G, H, Ψ, Φ_0) is a Lie quasi crossed module.

Next let $G_0 \subset G$ be the group of Lie algebra automorphisms of $\mathfrak{h}$. Then (G_0, H, Ψ, Φ_0) is a Lie crossed module.

5.2 Additive Feedback and Compatible Connections

Let (G, H, Ψ, Φ_0) be a Lie quasi crossed module. We write $\mathfrak{g} := \mathrm{Lie}(G)$ and $\mathfrak{h} := \mathrm{Lie}(H)$. Recall that the Lie algebra $\mathfrak{h}$ is the tangent space to H at the element 1. Hence the multiplicative twisting Ψ induces a linear action

$$\Psi_{\mathfrak{h}} : G \to \mathrm{GL}(\mathfrak{h}).$$

We see that from the Lie quasi crossed module (G, H, Ψ, Φ_0) we obtain a twisting setup $(G, H, \Psi_{\mathfrak{h}})$. As in Chapter 4, we call $\Psi_{\mathfrak{h}}$ the *additive twisting*.

Recall that $\mathrm{Hom}(\mathfrak{h}, \mathfrak{g})$ is the space of $\mathbb{R}$-linear maps $\mathfrak{h} \to \mathfrak{g}$, and an element

$$\phi \in \mathcal{O}_{\mathrm{pws}}(X) \otimes \mathrm{Hom}_{\mathbb{R}}(\mathfrak{h}, \mathfrak{g})$$

is called a piecewise smooth family of linear maps from $\mathfrak{h}$ to $\mathfrak{g}$.

Definition 5.2.1. Let (G, H, Ψ, Φ_0) be a Lie quasi crossed module, and let (X, x_0) be a pointed polyhedron. An *additive feedback for* (G, H, Ψ, Φ_0) *over* (X, x_0) is an element

$$\Phi_X \in \mathcal{O}_{\mathrm{pws}}(X) \otimes \mathrm{Hom}(\mathfrak{h}, \mathfrak{g})$$

satisfying this condition:

($**$) There is equality

$$\mathrm{Lie}(\Phi_0) = \Phi_X(x_0)$$

in $\mathrm{Hom}(\mathfrak{h}, \mathfrak{g})$.

Definition 5.2.2. Let (X, x_0) be a pointed polyhedron. A *Lie quasi crossed module with additive feedback over* (X, x_0) is the data

$$\mathbf{C}/X = (G, H, \Psi, \Phi_0, \Phi_X)$$

consisting of:

- A Lie quasi crossed module $\mathbf{C} = (G, H, \Psi, \Phi_0)$.
- An additive feedback Φ_X for $\mathbf{C}$ over (X, x_0).

When we talk about a Lie quasi crossed module with additive feedback $\mathbf{C}/X$, by default we use the notation of Definitions 5.2.1 and 5.2.2, and we write $\mathfrak{g} := \mathrm{Lie}(G)$ and $\mathfrak{h} := \mathrm{Lie}(H)$.

Let $\mathbf{C}/X$ be a Lie quasi crossed module with additive feedback over (X, x_0). Given a piecewise linear map $f : (Y, y_0) \to (X, x_0)$ between pointed polyhedra, consider

$$f^*(\Phi_X) \in \mathcal{O}_{\mathrm{pws}}(Y) \otimes \mathrm{Hom}_{\mathbb{R}}(\mathfrak{h}, \mathfrak{g}).$$

Then

(5.2.3) $$f^*(\mathbf{C}/X) := (G, H, \Psi, \Phi_0, f^*(\Phi_X))$$

is a Lie quasi crossed module with additive feedback over (Y, y_0).

Definition 5.2.4. Let $\mathbf{C}/X$ be a Lie quasi crossed module with additive feedback over (X, x_0). A *connection compatible with* $\mathbf{C}/X$ is a differential form

$$\alpha \in \Omega^1_{\mathrm{pws}}(X) \otimes \mathfrak{g}$$

satisfying the *holonomy condition*:

($\Diamond$) Let σ be a string in X, with $x_0 = \sigma(v_0)$ and $x_1 := \sigma(v_1)$. Define

$$g := \mathrm{MI}(\alpha \,|\, \sigma) \in G.$$

Then the diagram

$$
\begin{array}{ccc}
\mathfrak{h} & \xleftarrow{\ \Psi_{\mathfrak{h}}(g)\ } & \mathfrak{h} \\
{\scriptstyle \Phi_X(x_0)}\big\downarrow & & \big\downarrow{\scriptstyle \Phi_X(x_1)} \\
\mathfrak{g} & \xleftarrow{\ \mathrm{Ad}_{\mathfrak{g}}(g)\ } & \mathfrak{g}
\end{array}
$$

is commutative.

It could happen that $\mathbf{C}/X$ does not admit any compatible connection.

Example 5.2.5. Suppose (G, H, Ψ, Φ_0) is a Lie crossed module (see Remark 5.1.2) and (X, x_0) is a pointed polyhedron. Define $\Phi_X := \mathrm{Lie}(\Phi_0)$; this is a G-equivariant Lie algebra map $\mathfrak{h} \to \mathfrak{g}$, which we view as a constant element of $\mathcal{O}_{\mathrm{pws}}(X) \otimes \mathrm{Hom}(\mathfrak{h}, \mathfrak{g})$. In this way we obtain a Lie quasi crossed module with additive feedback $\mathbf{C}/X := (G, H, \Psi, \Phi_0, \Phi_X)$ over (X, x_0). Since Φ_X is G-equivariant, it follows that any $\alpha \in \Omega^1_{\mathrm{pws}}(X) \otimes \mathfrak{g}$ is a connection compatible with $\mathbf{C}/X$.

An example of an additive feedback, and of a compatible connection, is given in Section 5.6.

To a Lie quasi crossed module with additive feedback $\mathbf{C}/X$ there are two naturally associated twisting setups, namely $(G, H, \Psi_{\mathfrak{h}})$ and $(G, G, \mathrm{Ad}_{\mathfrak{g}})$.

Proposition 5.2.6. *Let*

$$\mathbf{C}/X = (G, H, \Psi, \Phi_0, \Phi_X)$$

be a Lie quasi crossed module with additive feedback over (X, x_0). *Then:*

(1) *The data* $\Theta_X := (\mathrm{id}_G, \Phi_0, \Phi_X)$ *is a transfer of twisting setups*

$$(G, H, \Psi_{\mathfrak{h}}) \to (G, G, \mathrm{Ad}_{\mathfrak{g}})$$

parametrized by (X, x_0), *in the sense of Definition 4.6.1.*

(2) *A form* $\alpha \in \Omega^1_{\mathrm{pws}}(X) \otimes \mathfrak{g}$ *is a connection compatible with* $\mathbf{C}/X$ *if and only if it is a connection compatible with* Θ_X, *in the sense of Definition 4.6.2.*

Proof. Immediate from the definitions.   □

Proposition 5.2.7. *Let* $\mathbf{C}/X$ *be a Lie quasi crossed module with additive feedback over* (X, x_0), *and let* α *be a compatible connection for* $\mathbf{C}/X$. *Suppose* $f : (Y, y_0) \to (X, x_0)$ *is a piecewise linear map between pointed polyhedra. Then* $f^*(\alpha)$ *is a compatible connection for* $f^*(\mathbf{C}/X)$.

Proof. Take a string σ in Y with $\sigma(v_0) = y_0$, and let $y_1 := \sigma(v_1)$. Then $f \circ \sigma$ is a string in X. According to Proposition 3.5.9 we have

$$\mathrm{MI}(\alpha \mid f \circ \sigma) = \mathrm{MI}(f^*(\alpha) \mid \sigma).$$

Let us call this element g. Let $x_1 := f(y_1)$. Then

$$\Phi_X(x_i) = f^*(\Phi_X)(y_i)$$

as homomorphisms $\mathfrak{h} \to \mathfrak{g}$, for $i = 0, 1$. We see that the holonomy condition is satisfied for $f^*(\alpha)$, relative to $f^*(\mathbf{C}/X)$.   □

Definition 5.2.8. Suppose

$$\mathbf{C}/X = (G, H, \Psi, \Phi_0, \Phi_X)$$

and

$$\mathbf{C}'/X = (G', H', \Psi', \Phi'_0, \Phi'_X)$$

are two Lie quasi crossed modules with additive feedbacks over (X, x_0). A *transfer* between them is a transfer of twisting setups

$$\Theta_X = (\Theta_G, \Theta_H, \Theta_{\mathfrak{h}, X}) : (G, H, \Psi_{\mathfrak{h}}) \to (G', H', \Psi'_{\mathfrak{h}'})$$

parametrized by (X, x_0), in the sense of Definition 4.6.1, satisfying this condition:

$$\Phi'_X \circ \Theta_{\mathfrak{h}, X} = \mathrm{Lie}(\Theta_G) \circ \Phi_X$$

in $\mathcal{O}_{\mathrm{pws}}(X) \otimes \mathrm{Hom}(\mathfrak{h}, \mathfrak{g}')$.

5.3   Connection-Curvature Pairs

Definition 5.3.1. Let $\mathbf{C}/X = (G, H, \Psi, \Phi_0, \Phi_X)$ be a Lie quasi crossed module with additive feedback over a pointed polyhedron (X, x_0); see Definition 5.2.2. A *connection-curvature pair* for $\mathbf{C}/X$ is a pair (α, β), consisting of Lie algebra valued differential forms

$$\alpha \in \Omega^1_{\mathrm{pws}}(X) \otimes \mathfrak{g}$$

and

$$\beta \in \Omega^2_{\mathrm{pws}}(X) \otimes \mathfrak{h},$$

satisfying the conditions below.

(i) α is a connection compatible with $\mathbf{C}/X$ (Definition 5.2.4).
(ii) The equation

$$\Phi_X(\beta) = \mathrm{d}(\alpha) + \tfrac{1}{2}[\alpha, \alpha]$$

holds in $\Omega^2_{\mathrm{pws}}(X) \otimes \mathfrak{g}$.

Remark 5.3.2. Condition (ii) above is often referred to as *vanishing of the fake curvature*. See [BM, BS].

Proposition 5.3.3. *Let $\mathbf{C}/X$ be a Lie quasi crossed module with additive feedback over (X, x_0), and let (α, β) be a connection-curvature pair for $\mathbf{C}/X$.*

(1) *Let $f : (Y, y_0) \to (X, x_0)$ be a piecewise linear map between pointed polyhedra. Then $(f^*(\alpha), f^*(\beta))$ is a connection-curvature pair in $f^*(\mathbf{C}/X)$.*

(2) *Let $\mathbf{C}'/X$ be another Lie quasi crossed module with additive feedback over (X, x_0), and let*

$$\Theta_X = (\Theta_G, \Theta_H, \Theta_{\mathfrak{h}, X}) : \mathbf{C}/X \to \mathbf{C}'/X$$

be a morphism of Lie quasi crossed module with additive feedback. Assume that α is compatible with $\Theta_{\mathfrak{h}, X}$. See Definitions 5.2.8 and 4.6.2. Then

$$\left(\mathrm{Lie}(\Theta_G)(\alpha), \Theta_{\mathfrak{h}, X}(\beta) \right)$$

is a connection-curvature pair for $\mathbf{C}'/X$.

Proof. (1) By Proposition 5.2.7 the form $\alpha' := f^*(\alpha)$ is a connection compatible with $f^*(\mathbf{C}/X)$. Next let us write $\beta' := f^*(\beta)$ and $\Phi'_X := f^*(\Phi_X)$. Since

$$f^* : \Omega^1_{\mathrm{pws}}(X) \otimes \mathfrak{g} \to \Omega^1_{\mathrm{pws}}(Y) \otimes \mathfrak{g}$$

is a DG Lie algebra homomorphism, we have

$$\Phi'_X(\beta') = f^*(\Phi_X(\beta)) = f^*\left(\mathrm{d}(\alpha) + \tfrac{1}{2}[\alpha, \alpha] \right) = \mathrm{d}(\alpha') + \tfrac{1}{2}[\alpha', \alpha'].$$

(2) The fact that α is compatible with $\Theta_{\mathfrak{h}, X}$, and that Θ_X satisfies the condition ($\Diamond$) in Definition 4.6.2, imply directly that $\alpha' := \mathrm{Lie}(\Theta_G)(\alpha)$ is a connection compatible with $\mathbf{C}'/X$.

Let $\beta' := \Theta_{\mathfrak{h}, X}(\beta)$. Then

$$\Phi'_X(\beta') = \mathrm{Lie}(\Theta_G)(\Phi_X(\beta)) = \mathrm{Lie}(\Theta_G)\left(\mathrm{d}(\alpha) + \tfrac{1}{2}[\alpha, \alpha] \right) = \mathrm{d}(\alpha') + \tfrac{1}{2}[\alpha', \alpha'].$$

We see that condition (ii) of Definition 5.3.1 holds. $\qquad\square$

Definition 5.3.4 (Tame Connection). Let $\mathbf{C}/X$ be a Lie quasi crossed module with additive feedback over (X, x_0). A form $\alpha \in \Omega^1_{\mathrm{pws}}(X) \otimes \mathfrak{g}$ is called a *tame connection* for $\mathbf{C}/X$ if there exists a form $\beta \in \Omega^2_{\mathrm{pws}}(X) \otimes \mathfrak{h}$ such that (α, β) is a connection-curvature pair (as in Definition 5.3.1).

In other words, α is a tame connection if it is a compatible connection (Definition 5.2.4), and its curvature $d(\alpha) + \frac{1}{2}[\alpha, \alpha]$ comes from $\Omega^2_{\mathrm{pws}}(X) \otimes \mathfrak{h}$.

Corollary 5.3.5. *In the situation of Proposition 5.3.3, the forms $f^*(\alpha)$ and $\mathrm{Lie}(\Theta_G)(\alpha)$ are tame connections.*

The proof is trivial.

5.4 Moving the Base Point

In this section we consider the following setup:

$$\mathbf{C}/X = (G, H, \Psi, \Phi_0, \Phi_X)$$

is a Lie quasi crossed module with additive feedback over a pointed polyhedron (X, x_0). We are given a form $\alpha \in \Omega^1_{\mathrm{pws}}(X) \otimes \mathfrak{g}$, which is a connection compatible with $\mathbf{C}/X$. And we are given a string ρ in X, with initial point $\rho(v_0) = x_0$ and terminal point $x_1 := \rho(v_1)$. Let

$$g := \mathrm{MI}(\alpha \,|\, \rho) \in G.$$

Recall that $\Psi(g)$ is an automorphism of the pointed analytic manifold $(H, 1)$. We define a new multiplication on the manifold H, by the formula

$$h_1 \cdot^g h_2 := \Psi(g)^{-1}\big(\Psi(g)(h_1) \cdot \Psi(g)(h_2)\big)$$

for $h_1, h_2 \in H$. In this way we obtain a new Lie group, that is denoted by H^g, and a Lie group isomorphism

$$\Psi(g) : H^g \to H.$$

See Figure 23.

The Lie algebra of H^g is $\mathfrak{h}^g$. So $\mathfrak{h}^g = \mathfrak{h}$ as vector spaces (this is the tangent space to H at 1), and

$$\Psi_{\mathfrak{h}}(g) : \mathfrak{h}^g \to \mathfrak{h}$$

is a Lie algebra isomorphism. There is a commutative diagram of maps

(5.4.1)

$$
\begin{array}{ccc}
\mathfrak{h} & \xleftarrow{\ \Psi_{\mathfrak{h}}(g)\ } & \mathfrak{h}^g \\
{\scriptstyle \exp_H} \downarrow & & \downarrow {\scriptstyle \exp_{H^g}} \\
H & \xleftarrow{\ \Psi(g)\ } & H^g
\end{array}
$$

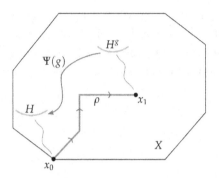

FIGURE 23. The string ρ from x_0 to x_1, the group element $g := \mathrm{MI}(\alpha \,|\, \rho) \in G$, and the group isomorphism $\Psi(g) : H^g \to H$.

Note that if $\Psi(g)$ is not a group automorphism of H, then H^g is not equal to H as groups. In this case the maps of manifolds

$$\exp_H, \exp_{H^g} : \mathfrak{h} \to H$$

could be distinct.

The data $(G, H^g, \Psi_{\mathfrak{h}})$ is a twisting setup (as in Definition 4.3.1), which in general is distinct from the twisting setup $(G, H, \Psi_{\mathfrak{h}})$, because of the possibly distinct exponential maps. Given a form $\beta \in \Omega^2_{\mathrm{pws}}(X) \otimes \mathfrak{h}$ and a kite (σ, τ) in (X, x_1), let us denote by

(5.4.2) $$\mathrm{MI}^g(\alpha, \beta \,|\, \sigma, \tau) \in H^g$$

the multiplicative integral with respect to the twisting setup $(G, H^g, \Psi_{\mathfrak{h}})$.

We define a map of Lie groups $\Phi_0^g : H^g \to G$ by the commutative diagram

$$
\begin{array}{ccc}
H & \xleftarrow{\ \Psi(g)\ } & H^g \\
{\scriptstyle \Phi_0} \downarrow & & \downarrow {\scriptstyle \Phi_0^g} \\
G & \xleftarrow{\ \mathrm{Ad}_G(g)\ } & G
\end{array}
$$

Proposition 5.4.3. *The data*

$$\mathbf{C}^g := (G, H^g, \Psi, \Phi_0^g)$$

is a Lie quasi crossed module.

Proof. Let us write $H' := H^g$, $G' := G$, $u := \mathrm{Ad}_G(g) : G' \to G$, $v := \Psi(g) : H' \to H$, $\Psi' := \Psi$ and $\Phi_0' := \Phi_0^g$. There are commutative diagrams of group homomorphisms

(5.4.4)

$$
\begin{array}{ccc}
H & \xleftarrow{\ v\ } & H' \\
\Phi_0 \downarrow & & \downarrow \Phi_0' \\
G & \xleftarrow{\ u\ } & G'
\end{array}
\qquad
\begin{array}{ccc}
G & \xleftarrow{\ u\ } & G' \\
\Psi \downarrow & & \downarrow \Psi' \\
\mathrm{Aut}(H) & \xleftarrow{\ \mathrm{Ad}(v)\ } & \mathrm{Aut}(H')
\end{array}
$$

Here $\mathrm{Aut}(H) = \mathrm{Aut}(H')$ is the group of automorphisms of the pointed manifold $H = H'$, and v is seen as an element of this group. The first diagram is just the definition of Φ_0'. The second diagram is commutative since for every $g' \in G'$ we have

$$
\begin{aligned}
\Psi(u(g')) &= \Psi(g \cdot g' \cdot g^{-1}) = \Psi(g) \cdot \Psi(g') \cdot \Psi(g^{-1}) \\
&= v \cdot \Psi(g') \cdot v^{-1} = (\mathrm{Ad}(v) \circ \Psi')(g').
\end{aligned}
$$

And by general group theory we have a commutative diagram

(5.4.5)

$$
\begin{array}{ccc}
H & \xleftarrow{\ v\ } & H' \\
\mathrm{Ad}_H \downarrow & & \downarrow \mathrm{Ad}_{H'} \\
\mathrm{Aut}(H) & \xleftarrow[\mathrm{Ad}(v)]{} & \mathrm{Aut}(H')
\end{array}
$$

We are given that $\Psi \circ \Phi_0 = \mathrm{Ad}_H$ (this is condition ($*$) of Definition 5.1.1 for the Lie quasi crossed module $\mathbf{C}$). Therefore by combining the three commutative diagrams we see that $\Psi' \circ \Phi_0' = \mathrm{Ad}_{H'}$. $\square$

Proposition 5.4.6. *The element Φ_X is an additive feedback for the Lie quasi crossed module $\mathbf{C}^g$ over the pointed polyhedron (X, x_1). Thus*

$$
\mathbf{C}^g / X := (G, H^g, \Psi, \Phi_0^g, \Phi_X)
$$

is a Lie quasi crossed module with additive feedback over (X, x_1).

Proof. In the notation used in the proof of the previous proposition, and with $\Phi_X' := \Phi_X$, we have to show that $\Phi_X'(x_1) = \mathrm{Lie}(\Phi_0')$, as linear maps $\mathfrak{h}' \to \mathfrak{g}'$. We have commutative diagrams of linear maps

$$
\begin{array}{ccc}
\mathfrak{h} & \xleftarrow{\ \mathrm{Lie}(v)\ } & \mathfrak{h}' \\
\mathrm{Lie}(\Phi_0) \downarrow & & \downarrow \mathrm{Lie}(\Phi_0') \\
\mathfrak{g} & \xleftarrow{\ \mathrm{Lie}(u)\ } & \mathfrak{g}'
\end{array}
\qquad
\begin{array}{ccc}
\mathfrak{h} & \xleftarrow{\ \Psi_{\mathfrak{h}}(g)\ } & \mathfrak{h} \\
\Phi_X(x_0) \downarrow & & \downarrow \Phi_X(x_1) \\
\mathfrak{g} & \xleftarrow[\mathrm{Ad}_{\mathfrak{g}}(g)]{} & \mathfrak{g}
\end{array}
$$

The first diagram is the differential of the first diagram in (5.4.4), and the second diagram is the holonomy condition for α relative to $\mathbf{C}/X$. Since Φ_X is an additive feedback for $\mathbf{C}$ over (X, x_0) we have $\Phi_X(x_0) = \mathrm{Lie}(\Phi_0)$.

We know that $\Psi_{\mathfrak{h}}(g) = \mathrm{Lie}(v)$ and $\mathrm{Ad}_{\mathfrak{g}}(g) = \mathrm{Lie}(u)$. It follows that $\Phi'_X(x_1) = \Phi_X(x_1) = \mathrm{Lie}(\Phi'_0)$. $\qquad\qquad\square$

Proposition 5.4.7. *The form α is a connection compatible with $\mathbf{C}^{g}/X$.*

Proof. We continue with the notation of the previous proofs. Let σ be a string in X initial point $\sigma(v_0) = x_1$ and terminal point $x_2 := \sigma(v_1)$. Define $g' := \mathrm{MI}(\alpha \,|\, \sigma) \in G' = G$. Because α is a connection compatible with $\mathbf{C}/X$, and because $g \cdot g' = \mathrm{MI}(\alpha \,|\, \rho * \sigma)$, we have a commutative diagram

$$
\begin{array}{ccccc}
\mathfrak{h} & \xrightarrow{\ \Psi_{\mathfrak{h}}(g)\ } & \mathfrak{h} & \xleftarrow{\ \Psi_{\mathfrak{h}}(g \cdot g')\ } & \mathfrak{h} \\
{\scriptstyle\Phi_X(x_1)}\big\downarrow & & {\scriptstyle\Phi_X(x_0)}\big\downarrow & & \big\downarrow{\scriptstyle\Phi_X(x_2)} \\
\mathfrak{g} & \xrightarrow[\ \mathrm{Ad}_{\mathfrak{g}}(g)\ ]{} & \mathfrak{g} & \xleftarrow[\ \mathrm{Ad}_{\mathfrak{g}}(g \cdot g')\ ]{} & \mathfrak{g}
\end{array}
$$

Since

$$\Psi_{\mathfrak{h}}(g)^{-1} \circ \Psi_{\mathfrak{h}}(g \cdot g') = \Psi_{\mathfrak{h}}(g')$$

and

$$\mathrm{Ad}_{\mathfrak{g}}(g)^{-1} \circ \mathrm{Ad}_{\mathfrak{g}}(g \cdot g') = \mathrm{Ad}_{\mathfrak{g}}(g')$$

we see that the diagram

$$
\begin{array}{ccc}
\mathfrak{h}' & \xleftarrow{\ \Psi'_{\mathfrak{h}'}(g')\ } & \mathfrak{h}' \\
{\scriptstyle\Phi'_X(x_1)}\big\downarrow & & \big\downarrow{\scriptstyle\Phi'_X(x_2)} \\
\mathfrak{g}' & \xleftarrow[\ \mathrm{Ad}_{\mathfrak{g}'}(g')\ ]{} & \mathfrak{g}'
\end{array}
$$

is commutative. $\qquad\qquad\square$

Theorem 5.4.8. *Let*

$$\mathbf{C}/X = (G, H, \Psi, \Phi_0, \Phi_X)$$

be a Lie quasi crossed module with additive feedback over a pointed polyhedron (X, x_0), let $\alpha \in \Omega^1_{\mathrm{pws}}(X) \otimes \mathfrak{g}$ be a connection compatible with $\mathbf{C}/X$, and let ρ be a string in X, with initial point $\rho(v_0) = x_0$. Define $g := \mathrm{MI}(\alpha \,|\, \rho)$, and let

$$\mathbf{C}^{g}/X = (G, H^{g}, \Psi, \Phi^{g}_0, \Phi_X)$$

be the Lie quasi crossed module with additive feedback over (X, x_1) constructed above.

    Given a form $\beta \in \Omega^2_{\mathrm{pws}}(X) \otimes \mathfrak{h}$ and a kite (σ, τ) in (X, x_1), consider the element

$$\mathrm{MI}^{g}(\alpha, \beta \,|\, \sigma, \tau) \in H^{g}$$

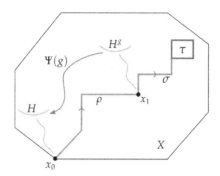

FIGURE 24. The string ρ from x_0 to x_1, and the kite (σ, τ) in (X, x_1).

from (5.4.2). *Then*

$$\Psi(g)\left(\mathrm{MI}^g(\alpha, \beta \mid \sigma, \tau)\right) = \mathrm{MI}(\alpha, \beta \mid \rho * \sigma, \tau)$$

in H.

See Figure 24.

Proof. We can assume that the kite (σ, τ) is nondegenerate. Define $Z := \tau(\mathbf{I}^2)$ and $g' := \mathrm{MI}^g(\alpha \mid \sigma)$. Choose a positively oriented orthonormal linear coordinate system (s_1, s_2) on Z.

Take $k \geq 0$ and $i \in \{1, \ldots, 4^k\}$, and define $z_i := (\tau \circ \tau_i^k)(\frac{1}{2}, \frac{1}{2}) \in Z$ and

$$g_i' := \mathrm{MI}\left((\alpha \mid (\tau \circ \sigma_i^k) * (\tau \circ \tau_i^k \circ \sigma_{\mathrm{pr}})\right).$$

So

$$\mathrm{MI}\left(\alpha \mid \rho * \sigma * (\tau \circ \sigma_i^k) * (\tau \circ \tau_i^k \circ \sigma_{\mathrm{pr}})\right) = g \cdot g' \cdot g_i'.$$

Assume that z_i is a smooth point of $\beta|_Z$. Let Y_i be a triangle in Z, and let $\tilde{\beta}_i \in \mathcal{O}(Y_i) \otimes \mathfrak{h}$, such that $z \in \mathrm{Int}\, Y_i$ and $\beta|_{Y_i} = \tilde{\beta}_i \cdot \mathrm{ds}_1 \wedge \mathrm{ds}_2$. According to Definition 4.3.3 and the commutative diagram (5.4.1) we have

$$\begin{aligned}
\mathrm{RP}_0&\left(\alpha, \beta \mid (\rho * \sigma, \tau) \circ (\sigma_i^k, \tau_i^k)\right) \\
&= \exp_H\left((\tfrac{1}{4})^k \cdot \mathrm{area}(Z) \cdot \Psi_{\mathfrak{h}}(g \cdot g' \cdot g_i')(\tilde{\beta}_i(z_i))\right) \\
&= \left(\Psi(g) \circ \exp_{H^g}\right)\left((\tfrac{1}{4})^k \cdot \mathrm{area}(Z) \cdot \Psi_{\mathfrak{h}}(g' \cdot g_i')(\tilde{\beta}_i(z_i))\right) \\
&= \Psi(g)\left(\mathrm{RP}_0^g(\alpha, \beta \mid (\sigma, \tau) \circ (\sigma_i^k, \tau_i^k))\right).
\end{aligned}$$

On the other hand, if z_i is a singular point of $\beta|_Z$ then

$$\mathrm{RP}_0\left(\alpha, \beta \mid (\rho * \sigma, \tau) \circ (\sigma_i^k, \tau_i^k)\right) = 1 = \Psi(g)\left(\mathrm{RP}_0^g(\alpha, \beta \mid (\sigma, \tau) \circ (\sigma_i^k, \tau_i^k))\right).$$

We see that

$$RP_k(\alpha, \beta \,|\, \rho * \sigma, \tau) = \Psi(g)\left(RP_k^g(\alpha, \beta \,|\, \sigma, \tau)\right)$$

for every k. Passing to the limit $k \to \infty$ finishes the proof.   □

5.5 Partial Differential Equations

Let $\alpha \in \Omega^1(\mathbf{I}^1) \otimes \mathfrak{g}$. For $x \in \mathbf{I}^1$ let $\sigma_x : \mathbf{I}^1 \to \mathbf{I}^1$ be the linear map defined on vertices by $\sigma_x(v_0, v_1) := (v_0, x)$. Let

$$(5.5.1) \qquad\qquad g(x) := \mathrm{MI}(\alpha \,|\, \sigma_x) \in G.$$

It is well known that the function $g : \mathbf{I}^1 \to G$ defined in this way is smooth, and moreover it satisfies the differential equation

$$(5.5.2) \qquad\qquad \mathrm{dlog}(g) = \alpha$$

with initial condition $g(0) = 1$. See [DF] for the case $G = \mathrm{GL}_n(\mathbb{R})$, in which

$$\mathrm{dlog}(g) = g^{-1} \cdot dg$$

as matrices. This ODE determines the function g.

Now consider a Lie quasi-crossed module $\mathbf{C}$, and a smooth connection-curvature pair (α, β) in $\mathbf{C}/\mathbf{I}^2$. For a point $x \in \mathbf{I}^2$ let $\tau_x : \mathbf{I}^2 \to \mathbf{I}^2$ be the linear map defined on vertices by

$$\tau_x(v_0, v_1, v_1) := \big(v_0, (t_1(x), 0), (0, t_2(x))\big).$$

And let σ be the empty string, so that (σ, τ_x) is a kite in $(\mathbf{I}^2, v_0)$. Define

$$h(x) := \mathrm{MI}(\alpha, \beta \,|\, \sigma, \tau_x) \in H.$$

Presumably the function $h : \mathbf{I}^2 \to H$ is smooth, and it satisfies a partial differential equation generalizing (5.5.2). We did not check this assertion.

In the very recent paper [BGNT] the authors consider a special case: the groups G and H are unipotent, $\mathbf{C}$ is a crossed module, and the pair (α, β) is algebraic. They write down a partial differential equation, whose unique solution is declared to be the multiplicative integral. Presumably this multiplicative integral coincides with ours. For more in this direction see Section 9.4.

5.6 Quantum Type DG Lie Algebras

Here we explain how Lie quasi crossed modules arise from deformation theory. This is continued in the subsequent section.

A *differential graded* (DG) Lie algebra is a graded $\mathbb{R}$-module $\mathfrak{f} = \bigoplus_{i \in \mathbb{Z}} \mathfrak{f}^i$ equipped with a graded Lie bracket $[-, -]$ and a differential d (of degree 1) that satisfy the graded Leibniz rule. For instance, if X is a manifold and

$\mathfrak{g}$ is a Lie algebra, then $\mathfrak{f} := \Omega(X) \otimes \mathfrak{g}$ is a DG Lie algebra. We say that $\mathfrak{f}$ is a *quantum type* DG Lie algebra if $\mathfrak{f}^i = 0$ for all $i < -1$; i.e. $\mathfrak{f} = \bigoplus_{i \geq -1} \mathfrak{f}^i$.

Let us fix a quantum type DG Lie algebra $\mathfrak{f}$, and assume that $\mathfrak{f}$ is nilpotent, and finite dimensional in each degree. (The assumptions of finiteness and nilpotence are for the sake of presentation; the construction works even when $\mathfrak{f}$ is infinite dimensional and pronilpotent.)

The Lie bracket $[-, -]$ of $\mathfrak{f}$ makes the vector space $\mathfrak{g} := \mathfrak{f}^0$ into a nilpotent Lie algebra. We denote by $G = \exp(\mathfrak{g})$ the corresponding unipotent group. In order to be concrete, we take G to be the analytic manifold $\mathfrak{g}$, made into a Lie group by the CBH formula (cf. Chapter 2); in particular $1_G := 0_{\mathfrak{g}}$. The exponential map $\exp_G : \mathfrak{g} \to G$ is just the identity map here.

The Lie algebra $\mathfrak{g}$ acts on the vector space $\mathfrak{f}^i$ (any i) by the adjoint action ad, namely

$$\mathrm{ad}(\alpha)(\beta) := [\alpha, \beta]$$

for $\alpha \in \mathfrak{g}$ and $\beta \in \mathfrak{f}^i$. Clearly $\mathrm{ad}(\alpha)$ is $\mathbb{R}$-linear. There is a second action of $\mathfrak{g}$ on $\mathfrak{f}^1$, which we call the *affine action*, and it is

$$\mathrm{af}(\alpha)(\beta) := \mathrm{d}(\alpha) - [\alpha, \beta]$$

for $\beta \in \mathfrak{f}^1$. The action af is usually not linear – it is an action by affine transformations. Both these actions integrate (or rather exponentiate) to actions Ad and Af of the group G on the vector spaces $\mathfrak{f}^i$ and $\mathfrak{f}^1$ respectively.

An element $\omega \in \mathfrak{f}^1$ is called an *MC element* if it satisfies the Maurer-Cartan equation

$$\mathrm{d}(\omega) + \tfrac{1}{2}[\omega, \omega] = 0.$$

We denote by $\mathrm{MC}(\mathfrak{f})$ the set of all MC elements. It turns out that the action Af of G on $\mathfrak{f}^1$ preserves the subset $\mathrm{MC}(\mathfrak{f})$. Namely if $\omega \in \mathrm{MC}(\mathfrak{f})$ and $g \in G$, then $\omega' := \mathrm{Af}(g)(\omega)$ is also in $\mathrm{MC}(\mathfrak{f})$. The quotient set by this action is denoted by $\overline{\mathrm{MC}}(\mathfrak{f})$.

Let us write $\mathfrak{h} := \mathfrak{f}^{-1}$. Say an MC element ω is given. Define an $\mathbb{R}$-linear map $\mathrm{d}_\omega : \mathfrak{h} \to \mathfrak{g}$ by the formula

(5.6.1) $\mathrm{d}_\omega(\beta) := \mathrm{d}(\beta) + [\omega, \beta]$

for $\beta \in \mathfrak{h}$. And define an $\mathbb{R}$-bilinear operation $[-, -]_\omega$ on $\mathfrak{h}$ by the formula

(5.6.2) $[\beta_1, \beta_2]_\omega := [\mathrm{d}_\omega(\beta_1), \beta_2].$

An elementary calculation shows that the vector space $\mathfrak{h}$ is a nilpotent Lie algebra with respect to the bracket $[-, -]_\omega$, and we denote this Lie algebra by $\mathfrak{h}_\omega$. A similar calculation shows that $\mathrm{d}_\omega : \mathfrak{h}_\omega \to \mathfrak{g}$ is a Lie algebra map. Let us define $H_\omega := \exp(\mathfrak{h}_\omega)$, the corresponding unipotent group, with multiplication $h_1 \cdot_\omega h_2$ and inverse $h_1^{-1_\omega}$, for $h_i \in H_\omega$. So as analytic manifolds we have $H_\omega = \mathfrak{h}_\omega = \mathfrak{h}$, and $\exp_{H_\omega} : \mathfrak{h}_\omega \to H_\omega$ is the identity map. There is a Lie group map $\exp(\mathrm{d}_\omega) : H_\omega \to G$. It is easy to see from (5.6.2) that for elements $h_1, h_2 \in H_\omega$ one has

$$\mathrm{Ad}\big(\exp(\mathrm{d}_\omega)(h_1)\big)(h_2) = \mathrm{Ad}_{H_\omega}(h_1)(h_2) = h_1 \cdot_\omega h_2 \cdot_\omega h_1^{-1_\omega}.$$

Suppose we are given $\omega \in \mathrm{MC}(\mathfrak{f})$ and $g \in G$. Define $\omega' := \mathrm{Af}(g)(\omega)$. A more difficult calculation shows that there is a commutative diagram of Lie algebra maps

$$\begin{array}{ccc} \mathfrak{h}_\omega & \xrightarrow{\ \mathrm{Ad}(g)\ } & \mathfrak{h}_{\omega'} \\ {\scriptstyle \mathrm{d}_\omega}\big\downarrow & & \big\downarrow{\scriptstyle \mathrm{d}_{\omega'}} \\ \mathfrak{g} & \xrightarrow[\ \mathrm{Ad}(g)\ ]{} & \mathfrak{g} \end{array}$$

See [Ye5, Section 1]. Hence there is also a commutative diagram of Lie group maps

$$\begin{array}{ccc} H_\omega & \xrightarrow{\ \exp(\mathrm{Ad}(g))\ } & H_{\omega'} \\ {\scriptstyle \exp(\mathrm{d}_\omega)}\big\downarrow & & \big\downarrow{\scriptstyle \exp(\mathrm{d}_{\omega'})} \\ G & \xrightarrow[\ \exp(\mathrm{Ad}(g))\ ]{} & G \end{array}$$

Observe that if $[-,-]_\omega \neq [-,-]_{\omega'}$ as Lie brackets, then $\mathfrak{h}_\omega \neq \mathfrak{h}_{\omega'}$ as Lie algebras, $H_\omega \neq H_{\omega'}$ as Lie groups, and $\Psi(g) := \exp(\mathrm{Ad}(g))$ is not a Lie group automorphism of H_ω; it is only an automorphism of pointed analytic manifolds. At least we obtain in this way an action Ψ of G on H_ω by automorphisms of pointed analytic manifolds.

Take an MC element ω, and define the map of Lie groups

$$\Phi_\omega := \exp(\mathrm{d}_\omega) : H_\omega \to G.$$

What we have said so far implies that the data

(5.6.3) $$\mathbf{C}_\omega := \big(G, H_\omega, \Psi, \Phi_\omega\big)$$

is a Lie quasi crossed module.

5.7 Cosimplicial DG Lie Algebras

Here we show how deformation quantization gives rise to Lie quasi crossed modules with additive feedback and connection-curvature pairs.

Suppose we are given a *cosimplicial nilpotent quantum type DG Lie algebra* $\mathfrak{f}$. So $\mathfrak{f} = \{\mathfrak{f}^{p,\cdot}\}_{p \in \mathbb{N}}$, where each $\mathfrak{f}^{p,\cdot} = \bigoplus_{i \geq -1} \mathfrak{f}^{p,i}$ is a nilpotent quantum type DG Lie algebra.

From the cosimplicial object $\mathfrak{f}$ one constructs its Thom-Sullivan normalization $\tilde{N}(\mathfrak{f})$, which is a nilpotent quantum type DG Lie algebra, involving algebraic differential forms on all the simplices Δ^p, $p \geq 0$. For details see [Ye2, Section 2].

Suppose $\tilde{\omega}$ is an MC element of $\tilde{N}(\mathfrak{f})$. By definition $\tilde{\omega} = \{\tilde{\omega}^{p,q}\}$, where

$$\tilde{\omega}^{p,q} \in \Omega^q(\Delta^p) \otimes \mathfrak{f}^{p,1-q}, \quad p \in \mathbb{N}, \ q \in \{0,1,2\}.$$

Let us fix a number p. We look at the components of $\tilde{\omega}$ with index p:

$$\omega := \tilde{\omega}^{p,0} \in \mathcal{O}(\Delta^p) \otimes \mathfrak{f}^{p,1},$$
$$\alpha := \tilde{\omega}^{p,1} \in \Omega^1(\Delta^p) \otimes \mathfrak{f}^{p,0}$$

and

$$\beta := \tilde{\omega}^{p,2} \in \Omega^2(\Delta^p) \otimes \mathfrak{f}^{p,-1}.$$

Now let us write $\mathfrak{g} := \mathfrak{f}^{p,0}$, $\mathfrak{h} := \mathfrak{f}^{p,-1}$ and $X := \Delta^p$. With this notation $\omega \in \mathcal{O}(X) \otimes \mathfrak{f}^{p,1}$, $\alpha \in \Omega^1(X) \otimes \mathfrak{g}$ and $\beta \in \Omega^2(X) \otimes \mathfrak{h}$.

Choose some point $x_0 \in X$. The element $\omega(x_0) \in \mathfrak{f}^{p,1}$ turns out to be an MC element of the DG Lie algebra $\mathfrak{f}^{p,\cdot}$. As in (5.6.3), we have a Lie quasi crossed module

$$\mathbf{C}_{\omega(x_0)} = \big(G, H_{\omega(x_0)}, \Psi, \Phi_{\omega(x_0)}\big).$$

Next we define

$$\Phi_X := \mathrm{d}_\omega \in \mathcal{O}(X) \otimes \mathrm{Hom}(\mathfrak{h}, \mathfrak{g});$$

see (5.6.1). This is an additive feedback for $\mathbf{C}_{\omega(x_0)}$ over the pointed polyhedron (X, x_0), so we obtain a Lie quasi crossed module with additive feedback

(5.7.1) $$\mathbf{C}_{\omega(x_0)} / X = \big(G, H_{\omega(x_0)}, \Psi, \Phi_{\omega(x_0)}, \Phi_X\big).$$

Furthermore, it can be shown that the pair (α, β) is a connection-curvature pair in $\mathbf{C}_{\omega(x_0)} / X$. Taking any kite (σ, τ) in (X, x_0) we thus obtain a group element

$$\mathrm{MI}(\alpha, \beta \,|\, \sigma, \tau) \in H_{\omega(x_0)}.$$

In Section 9.5 we continue with this setup, and state Conjecture 9.5.2.

6 Stokes Theorem in Dimension Two

The purpose of this chapter is to prove Theorem 6.2.1. When $H = G$ (see Example 6.2.7) this is just a fancy version of Schlesinger's theorem.

Let (X, x_0) be a pointed polyhedron. Recall the notions of Lie quasi crossed module with additive feedback

(6.0.1) $$\mathbf{C}/X = (G, H, \Psi, \Phi_0, \Phi_X)$$

over (X, x_0), connection-curvature pair (α, β), and multiplicative integral $\mathrm{MI}(\alpha, \beta \mid \sigma, \tau)$; see Definitions 5.2.2, 5.3.1 and 4.4.16 respectively.

6.1 Some Estimates

In this section we assume that $(X, x_0) = (\mathbf{I}^2, v_0)$. We fix a Lie quasi crossed module with additive feedback $\mathbf{C}/X$ over $(\mathbf{I}^2, v_0)$, in which $G = H$, as in Example 5.1.4. We also fix a connection-curvature pair (α, β) for $\mathbf{C}/X$. Note that the equality

(6.1.1) $$\beta = \mathrm{d}(\alpha) + \tfrac{1}{2}[\alpha, \alpha]$$

holds in $\Omega_{\mathrm{pws}}^2(\mathbf{I}^2) \otimes \mathfrak{g}$. As in Chapter 1 we choose a euclidean norm $\|-\|_{\mathfrak{g}}$ on the vector space $\mathfrak{g}$; an open neighborhood $V_0(G)$ of 1 in G on which $\log_G$ is well-defined; a convergence radius $\epsilon_0(G)$; and a commutativity constant $c_0(G)$.

In Proposition 4.5.4 we established certain constants $c_{2'}(\alpha, \beta)$ and $\epsilon_{2'}(\alpha, \beta)$.

Lemma 6.1.2. *There are constants $c_3(\alpha, \beta)$ and $\epsilon_3(\alpha, \beta)$ with these properties:*

(1) $c_3(\alpha, \beta) \geq c_{2'}(\alpha, \beta)$ *and* $0 < \epsilon_3(\alpha, \beta) \leq \epsilon_{2'}(\alpha, \beta)$.
(2) *Suppose (σ, τ) is a square kite in $(\mathbf{I}^2, v_0)$ such that* $\mathrm{side}(\tau) < \epsilon_3(\alpha, \beta)$ *and* $\mathrm{len}(\sigma) \leq 5$. *Then*

$$\mathrm{MI}(\alpha, \beta \mid \sigma, \tau), \ \mathrm{MI}(\alpha \mid \partial(\sigma, \tau)) \in V_0(G)$$

and

$$\left\| \log_G\big(\mathrm{MI}(\alpha, \beta \mid \sigma, \tau)\big) \right\|, \ \left\| \log_G\big(\mathrm{MI}(\alpha \mid \partial(\sigma, \tau))\big) \right\| \leq c_3(\alpha, \beta) \cdot \mathrm{side}(\tau)^2.$$

Proof. For $\mathrm{MI}(\alpha, \beta \mid \sigma, \tau)$ we can use the estimate from Proposition 4.5.2.

For the boundary we have to do some work. Let us write $\epsilon := \mathrm{side}(\tau)$, and suppose that $\epsilon < \tfrac{1}{4}\epsilon_1(\alpha)$, where $\epsilon_1(\alpha)$ is the constant from Definition

3.3.16. Consider the closed string $\partial\tau$ of length 4ϵ. Write $g_1 := \mathrm{MI}(\alpha \mid \partial\tau) \in G$. According to Proposition 3.5.11 we know that $g_1 \in V_0(G)$, and

$$\left\| \log_G(g_1) - \int_{\partial\tau} \alpha \right\| \leq c_0(G) \cdot c_1(\alpha)^2 \cdot (4\epsilon)^2 .$$

By the abelian Stokes Theorem (Theorem 1.8.3) we have

$$\int_{\partial\tau} \alpha = \int_\tau \mathrm{d}(\alpha).$$

Now

$$\left\| \int_\tau \mathrm{d}(\alpha) \right\| \leq \mathrm{area}(\tau) \cdot \|\alpha\|_{\mathrm{Sob}} = \epsilon^2 \cdot \|\alpha\|_{\mathrm{Sob}} .$$

We conclude that

$$\|\log_G(g_1)\| \leq \left(c_0(G) \cdot c_1(\alpha)^2 \cdot 16 + \|\alpha\|_{\mathrm{Sob}} \right) \cdot \epsilon^2 .$$

Next let $g_2 := \mathrm{MI}(\alpha \mid \sigma) \in G$. Consider the representation $\mathrm{Ad}_\mathfrak{g} : G \to \mathrm{GL}(\mathfrak{g})$. By Proposition 3.5.10 the norm of the operator $\mathrm{Ad}_\mathfrak{g}(g_2)$ satisfies

$$\|\mathrm{Ad}_\mathfrak{g}(g_2)\| \leq \exp(c_4(\alpha, \mathrm{Ad}_\mathfrak{g}) \cdot 5) .$$

Finally we look at $g := \mathrm{MI}(\alpha \mid \partial(\sigma, \tau))$. By definition

$$g = g_2 \cdot g_1 \cdot g_2^{-1} = \mathrm{Ad}_G(g_2)(g_1).$$

The logarithm is

$$\log_G(g) = \mathrm{Ad}_\mathfrak{g}(g_2)(\log_G(g_1)).$$

By combining the estimates above we get

$$\|\log_G(g)\| \leq \exp(c_4(\alpha, \mathrm{Ad}_\mathfrak{g}) \cdot 5) \cdot \left(c_0(G) \cdot c_1(\alpha)^2 \cdot 16 + \|\alpha\|_{\mathrm{Sob}} \right) \cdot \epsilon^2 .$$

We see that we can take the following constants:

$$\epsilon_3(\alpha, \beta) := \min\left(\epsilon_2(\alpha, \beta), \tfrac{1}{4}\epsilon_1(\alpha) \right)$$

and

$$c_4(\alpha, \beta) :=$$
$$\max\left(c_2(\alpha, \beta), \exp(c_4(\alpha, \mathrm{Ad}_\mathfrak{g}) \cdot 5) \cdot \left(c_0(G) \cdot c_1(\alpha)^2 \cdot 16 + \|\alpha\|_{\mathrm{Sob}} \right) \right) .$$

$\square$

Lemma 6.1.3. *There are constants $c_4(\alpha, \beta)$ and $\epsilon_4(\alpha, \beta)$ with these properties:*

(1) $c_3(\alpha, \beta) \leq c_4(\alpha, \beta)$ and $0 < \epsilon_4(\alpha, \beta) \leq \epsilon_3(\alpha, \beta)$.

(2) *Suppose (σ, τ) is a square kite in $(\mathbf{I}^2, v_0)$ such that $\mathrm{side}(\tau) < \epsilon_4(\alpha, \beta)$, $\mathrm{len}(\sigma) \leq 5$, and $\alpha|_{\tau(\mathbf{I}^2)}$ is smooth. Then*

$$\left\| \log_G\left(\mathrm{MI}(\alpha, \beta \mid \sigma, \tau) \right) - \log_G\left(\mathrm{MI}(\alpha \mid \partial(\sigma, \tau)) \right) \right\| \leq c_4(\alpha, \beta) \cdot \mathrm{side}(\tau)^3 .$$

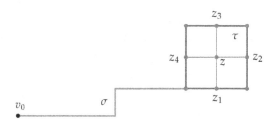

FIGURE 25. Computing $RP_0(\alpha \,|\, \partial\tau)$ for a tiny square kite (σ, τ) in $(\mathbf{I}^2, v_0)$.

Proof. Let $\epsilon := \mathrm{side}(\tau)$, $Z := \tau(\mathbf{I}^2)$, and $z := \tau(\tfrac{1}{2}, \tfrac{1}{2})$, which is the midpoint of the square Z. Denote by $(\rho_1, \ldots, \rho_4)$ the closed string $\partial\tau$. For any $i \in \{1, \ldots, 4\}$ let z_i be the midpoint of the edge $\rho_i(\mathbf{I}^1)$. See Figure 25 for an illustration. Let (s_1, s_2) be the positively oriented orthonormal linear coordinate system on Z, such that

$$\tau^*(s_i) = \epsilon \cdot (t_i - \tfrac{1}{2}).$$

So in particular $s_i(z) = 0$.

Since $\alpha|_Z$ is smooth, there are functions $\tilde{\alpha}_1, \tilde{\alpha}_2 \in \mathcal{O}(Z) \otimes \mathfrak{g}$ such that

$$\alpha|_Z = \tilde{\alpha}_1 \cdot ds_1 + \tilde{\alpha}_2 \cdot ds_2.$$

The Taylor expansion of $\tilde{\alpha}_j$ around z to second order looks like this:
(6.1.4)
$$\tilde{\alpha}_j(x) = \tilde{\alpha}_j(z) + \sum_{1 \le k \le 2} \left(\tfrac{\partial}{\partial s_k} \tilde{\alpha}_j \right)(z) \cdot s_k(x) + \sum_{1 \le k, l \le 2} g_{j,k,l}(x) \cdot s_k(x) \cdot s_l(x)$$

for $x \in Z$. Here $g_{j,k,l} : Z \to \mathfrak{g}$ are continuous functions satisfying

(6.1.5) $$\|g_{j,k,l}\| \le \|\alpha\|_{\mathrm{Sob}}.$$

We define elements $\lambda_i \in \mathfrak{g}$ as follows:

(6.1.6)
$$\begin{aligned}
\lambda_1 &:= \epsilon \cdot \tilde{\alpha}_1(z_1), \\
\lambda_2 &:= \epsilon \cdot \tilde{\alpha}_2(z_2), \\
\lambda_3 &:= -\epsilon \cdot \tilde{\alpha}_1(z_3), \\
\lambda_4 &:= -\epsilon \cdot \tilde{\alpha}_2(z_4).
\end{aligned}$$

Then, almost by definition,

$$RP_0(\alpha \,|\, \rho_i) = \exp_G(\lambda_i)$$

and

(6.1.7) $$\mathrm{RP}_0(\alpha \mid \partial \tau) = \prod_{i=1}^{4} \exp_G(\lambda_i).$$

Now

$$s_k(z_i) \in \{0, \tfrac{1}{2}\epsilon, -\tfrac{1}{2}\epsilon\},$$

and the value 0 occurs half the time. So the Taylor expansion (6.1.4) for the point z_i has only one summand of order 1 in ϵ (instead of two). Let us define

$$\mu_1 := \epsilon \cdot \tilde{\alpha}_1(z), \quad \mu_1' := -\tfrac{1}{2}\epsilon^2 \cdot \left(\tfrac{\partial}{\partial s_2}\tilde{\alpha}_1\right)(z),$$
$$\mu_2 := \epsilon \cdot \tilde{\alpha}_2(z), \quad \mu_2' := \tfrac{1}{2}\epsilon^2 \cdot \left(\tfrac{\partial}{\partial s_1}\tilde{\alpha}_2\right)(z),$$
$$\mu_3 := -\epsilon \cdot \tilde{\alpha}_1(z), \quad \mu_3' := -\tfrac{1}{2}\epsilon^2 \cdot \left(\tfrac{\partial}{\partial s_2}\tilde{\alpha}_1\right)(z),$$
$$\mu_4 := -\epsilon \cdot \tilde{\alpha}_2(z), \quad \mu_4' := \tfrac{1}{2}\epsilon^2 \cdot \left(\tfrac{\partial}{\partial s_1}\tilde{\alpha}_2\right)(z).$$

Then using the estimate (6.1.5) for the quadratic terms in (6.1.4) we obtain

(6.1.8) $$\| \lambda_i - (\mu_i + \mu_i') \| \leq \epsilon^3 \cdot \|\alpha\|_{\mathrm{Sob}}.$$

We also have these bounds:

(6.1.9) $$\|\lambda_i\| \leq \epsilon \cdot \|\alpha\|_{\mathrm{Sob}}$$

and

(6.1.10) $$\|\mu_i\| \leq \epsilon \cdot \|\alpha\|_{\mathrm{Sob}}, \quad \|\mu_i'\| \leq \tfrac{1}{2}\epsilon^2 \cdot \|\alpha\|_{\mathrm{Sob}}.$$

According to property (iv) of Theorem 2.1.2, the estimates (6.1.8) and (6.1.9) yield

(6.1.11)
$$\left\| \log_G\left(\prod_{i=1}^{4} \exp_G(\lambda_i)\right) - \log_G\left(\prod_{i=1}^{4} \exp_G(\mu_i + \mu_i')\right) \right\|$$
$$\leq 4\epsilon^3 \cdot c_0(G) \cdot \|\alpha\|_{\mathrm{Sob}}$$

for sufficiently small ϵ. Similarly the estimates (6.1.10) give us

(6.1.12)
$$\left\| \log_G\left(\prod_{i=1}^{4} \exp_G(\mu_i + \mu_i')\right) - \log_G\left(\prod_{i=1}^{4} \exp_G(\mu_i)\right) - \sum_{i=1}^{4} \mu_i' \right\|$$
$$\leq 2\epsilon^3 \cdot c_0(G) \cdot \|\alpha\|_{\mathrm{Sob}}^2.$$

Now $\mu_3 = -\mu_1$ and $\mu_4 = -\mu_2$, and hence

$$\prod_{i=1}^{4} \exp_G(\mu_i) = \exp_G(\mu_1) \cdot \exp_G(\mu_2) \cdot \exp_G(-\mu_1) \cdot \exp_G(-\mu_2).$$

According to property (iii) of Theorem 2.1.2 we see that

(6.1.13) $$\left\| \log_G\left(\prod_{i=1}^{4} \exp_G(\mu_i)\right) - [\mu_1, \mu_2] \right\| \leq 2\epsilon^3 \cdot c_0(G) \cdot \|\alpha\|_{\mathrm{Sob}}^3.$$

For the terms μ_i' we have

(6.1.14) $$\sum_{i=1}^{4} \mu_i' = -\epsilon^2 \cdot (\tfrac{\partial}{\partial s_2}\tilde{\alpha}_1)(z) + \epsilon^2 \cdot (\tfrac{\partial}{\partial s_1}\tilde{\alpha}_2)(z).$$

Putting together equations (6.1.7), (6.1.11), (6.1.12), (6.1.13) and (6.1.14) we conclude that for some $c \geq 1$ (depending on (α, β)) the estimate

$$\| \log_G (\mathrm{RP}_0(\alpha \,|\, \partial\tau)) - \epsilon^2 \cdot ([\tilde{\alpha}_1(z), \tilde{\alpha}_2(z)] - (\tfrac{\partial}{\partial s_2}\tilde{\alpha}_1)(z) + (\tfrac{\partial}{\partial s_1}\tilde{\alpha}_2)(z)) \|$$
$$\leq \epsilon^3 \cdot c$$

holds for sufficiently small ϵ. Using Proposition 3.3.22(2) we get
(6.1.15)
$$\| \log_G (\mathrm{MI}(\alpha \,|\, \partial\tau)) - \epsilon^2 \cdot ([\tilde{\alpha}_1(z), \tilde{\alpha}_2(z)] - (\tfrac{\partial}{\partial s_2}\tilde{\alpha}_1)(z) + (\tfrac{\partial}{\partial s_1}\tilde{\alpha}_2)(z)) \|$$
$$\leq \epsilon^3 \cdot c'$$

for a suitable constant c'.

Since equation (6.1.1) holds, we know that β is smooth on Z. Let $\tilde{\beta} \in \mathcal{O}(Z) \otimes \mathfrak{g}$ be such that $\beta|_Z = \tilde{\beta} \cdot \mathrm{d}s_1 \wedge \mathrm{d}s_2$. Then (6.1.1) becomes

$$\tilde{\beta} = [\tilde{\alpha}_1, \tilde{\alpha}_2] - (\tfrac{\partial}{\partial s_2}\tilde{\alpha}_1) + (\tfrac{\partial}{\partial s_1}\tilde{\alpha}_2),$$

as smooth functions $Z \to \mathfrak{g}$. Thus we can rewrite (6.1.15) as

(6.1.16) $$\| \log_G (\mathrm{MI}(\alpha \,|\, \partial\tau)) - \epsilon^2 \cdot \tilde{\beta}(z)) \| \leq \epsilon^3 \cdot c'.$$

Letting $g := \mathrm{MI}(\alpha \,|\, \sigma)$, we have (by Proposition 3.5.8):

$$\mathrm{MI}(\alpha \,|\, \partial(\sigma, \tau)) = g \cdot \mathrm{MI}(\alpha \,|\, \partial\tau) \cdot g^{-1} = \mathrm{Ad}_G(g)(\mathrm{MI}(\alpha \,|\, \partial\tau)).$$

The map $\log_G$ sends $\mathrm{Ad}_G(g)$ to $\mathrm{Ad}_{\mathfrak{g}}(g)$, and therefore

$$\log_G \big(\mathrm{MI}(\alpha \,|\, \partial(\sigma, \tau))\big) = \mathrm{Ad}_{\mathfrak{g}}(g)\big(\log_G (\mathrm{MI}(\alpha \,|\, \partial\tau))\big).$$

Recall that the length of σ is bounded by 5, so by Proposition 3.4.3 the norm of the operator $\mathrm{Ad}_{\mathfrak{g}}(g)$ on $\mathfrak{g}$ is also bounded. Plugging in the estimate (6.1.16) we now arrive at

(6.1.17) $$\| \log_G (\mathrm{MI}(\alpha \,|\, \partial(\sigma, \tau))) - \epsilon^2 \cdot \mathrm{Ad}_{\mathfrak{g}}(g)(\tilde{\beta}(z)) \| \leq \epsilon^3 \cdot c'''$$

for a suitable bound c''', again depending on (α, β).

Finally, by definition we have

$$\log_G (\mathrm{RP}_0(\alpha, \beta \,|\, \sigma, \tau)) = \epsilon^2 \cdot \mathrm{Ad}_{\mathfrak{g}}(g)(\tilde{\beta}(z)).$$

According to Proposition 4.5.2(2) we know that

$$\| \log_G (\mathrm{RP}_0(\alpha, \beta \,|\, \sigma, \tau)) - \log_G (\mathrm{MI}(\alpha, \beta \,|\, \sigma, \tau)) \| \leq c_2(\alpha, \beta) \cdot \epsilon^4.$$

Combining this with (6.1.17) we get

$$\big\| \log_G \big(\mathrm{MI}(\alpha, \beta \,|\, \sigma, \tau) \big) - \log_G \big(\mathrm{MI}(\alpha \,|\, \partial(\sigma, \tau)) \big) \big\| \leq c_4(\alpha, \beta) \cdot \epsilon^3$$

for a sufficiently large constant $c_4(\alpha, \beta)$ and for all sufficiently small ϵ. This gives us a value for $\epsilon_4(\alpha, \beta)$. $\qquad\square$

Definition 6.1.18. Let us fix constants $c_4(\alpha, \beta)$ and $\epsilon_4(\alpha, \beta)$ as in Lemma 6.1.3. A square kite (σ, τ) in $(\mathbf{I}^2, v_0)$ will be called (α, β)-*tiny* in this chapter if $\mathrm{side}(\tau) < \epsilon_4(\alpha, \beta)$ and $\mathrm{len}(\sigma) \leq 5$.

Lemma 6.1.19. *Let (σ, τ) be a kite in $(\mathbf{I}^2, v_0)$. Take some $k \geq 0$. For $i \in \{1, \ldots, 4^k\}$ let*

$$(\sigma_i, \tau_i) := (\sigma, \tau) \circ (\sigma_i^k, \tau_i^k) = \mathrm{tes}_i^k(\sigma, \tau).$$

Then

$$\prod_{i=1}^{4^k} \mathrm{MI}\big(\alpha \,|\, \partial(\sigma_i, \tau_i)\big) = \mathrm{MI}\big(\alpha \,|\, \partial(\sigma, \tau)\big).$$

Proof. By Proposition 3.5.8 we get cancellation of the contribution of all inner edges. $\qquad\square$

6.2   Stokes Theorem

Theorem 6.2.1 (Nonabelian Stokes Theorem in Dimension 2)*. Let (X, x_0) be a pointed polyhedron, let $\mathbf{C}/X$ be a Lie quasi crossed module with additive feedback over (X, x_0), and let (α, β) be a connection-curvature pair for $\mathbf{C}/X$. Then for any kite (σ, τ) in (X, x_0) one has*

$$\Phi_0\big(\mathrm{MI}(\alpha, \beta \,|\, \sigma, \tau) \big) = \mathrm{MI}\big(\alpha \,|\, \partial(\sigma, \tau) \big)$$

in G.

Proof. According to Proposition 5.2.6(1) there is a transfer of twisting setups

$$(\mathrm{id}_G, \Phi_0, \Phi_X) : (G, H, \Psi_\mathfrak{h}) \to (G, G, \mathrm{Ad}_\mathfrak{g})$$

parametrized by (X, x_0). Hence by Propositions 4.6.5 and 5.3.3(2) we can assume that $G = H$. Next, using Propositions 4.1.6, 4.5.3 and 5.3.3(1) we can further assume that $(X, x_0) = (\mathbf{I}^2, v_0)$, (σ, τ) is a square kite, and $\mathrm{len}(\sigma) \leq 1$. We need to prove that

(6.2.2) $$\mathrm{MI}(\alpha, \beta \,|\, \sigma, \tau) = \mathrm{MI}(\alpha \,|\, \partial(\sigma, \tau))$$

in G.

Take k large enough such that all the kites

(6.2.3) $$(\sigma_i, \tau_i) := \mathrm{tes}_i^k(\sigma, \tau)$$

in the k-th binary tessellation of (σ, τ) are (α, β)-tiny. By Proposition 4.5.1 we have

$$\mathrm{MI}(\alpha, \beta \,|\, \sigma, \tau) = \prod_{i=1}^{4^k} \mathrm{MI}(\alpha, \beta \,|\, \sigma_i, \tau_i).$$

Using Lemma 6.1.19 we see that it suffices to prove that

$$\mathrm{MI}(\alpha, \beta \,|\, \sigma_i, \tau_i) = \mathrm{MI}(\alpha \,|\, \partial(\sigma_i, \tau_i))$$

for every i. In this way we have reduced the problem to proving that

(6.2.4) $$\mathrm{MI}(\alpha, \beta \,|\, \sigma, \tau) = \mathrm{MI}(\alpha \,|\, \partial(\sigma, \tau))$$

for any (α, β)-tiny kite (σ, τ) in $(\mathbf{I}^2, v_0)$.

So assume (σ, τ) is (α, β)-tiny, with $\epsilon := \mathrm{side}(\tau)$. Take some $k \geq 0$, and let (σ_i, τ_i) be like in (6.2.3). Then $\mathrm{side}(\tau_i) = (\frac{1}{2})^k \cdot \epsilon$. We know that

(6.2.5) $$\mathrm{MI}(\alpha, \beta \,|\, \sigma, \tau) = \prod_{i=1}^{4^k} \mathrm{MI}(\alpha, \beta \,|\, \sigma_i, \tau_i)$$

and

(6.2.6) $$\mathrm{MI}(\alpha \,|\, \partial(\sigma, \tau)) = \prod_{i=1}^{4^k} \mathrm{MI}(\alpha \,|\, \partial(\sigma_i, \tau_i)).$$

Let us do some estimates now. If i is a bad index (in the sense of Definition 4.4.12), then by Lemma 6.1.2 we know that

$$\left\| \log_G \left(\mathrm{MI}(\alpha, \beta \,|\, \sigma_i, \tau_i) \right) \right\| \leq c_3(\alpha, \beta) \cdot (\tfrac{1}{2})^{2k} \epsilon^2$$

and

$$\left\| \log_G \left(\mathrm{MI}(\alpha \,|\, \partial(\sigma_i, \tau_i)) \right) \right\| \leq c_3(\alpha, \beta) \cdot (\tfrac{1}{2})^{2k} \epsilon^2.$$

On the other hand, if i is a good index, then Lemma 6.1.3 says that

$$\left\| \log_G \left(\mathrm{MI}(\alpha, \beta \,|\, \sigma_i, \tau_i) \right) - \log_G \left(\mathrm{MI}(\alpha \,|\, \partial(\sigma_i, \tau_i)) \right) \right\| \leq c_4(\alpha, \beta) \cdot (\tfrac{1}{2})^{3k} \epsilon^3.$$

Using these estimates, equations (6.2.5)-(6.2.6), Lemma 4.4.13 and property (iv) of Theorem 2.1.2, we arrive at

$$\left\| \log_G \left(\mathrm{MI}(\alpha, \beta \,|\, \sigma, \tau) \right) - \log_G \left(\mathrm{MI}(\alpha \,|\, \partial(\sigma, \tau)) \right) \right\|$$
$$\leq \underbrace{4^k \cdot c_4(\alpha, \beta) \cdot (\tfrac{1}{2})^{3k} \epsilon^3}_{\text{good indices}} + \underbrace{(a_0 + a_1 \cdot 2^k) \cdot c_3(\alpha, \beta) \cdot (\tfrac{1}{2})^{2k} \epsilon^2}_{\text{bad indices}}.$$

Here $a_i := a_i(\alpha, \beta, \mathbf{I}^2)$ are the constants from Lemma 4.4.13. Since the right hand side of this inequality tends to 0 as $k \to \infty$, we conclude that (6.2.4) holds. $\qquad \square$

Example 6.2.7. A special case of the corollary is the situation of Example 5.1.4. Take any differential form $\alpha \in \Omega^1_{\text{pws}}(X) \otimes \mathfrak{g}$, and let $\beta := d(\alpha) + \frac{1}{2}[\alpha, \alpha]$. Then (α, β) is a connection-curvature pair, and the corollary says that

$$\text{MI}(\alpha, \beta \mid \sigma, \tau) = \text{MI}(\alpha \mid \partial(\sigma, \tau)).$$

This is just Schlesinger's theorem [DF].

6.3　The Fundamental Relation

Here is an important consequence of Theorem 6.2.1.

Theorem 6.3.1 (The Fundamental Relation). *Let (X, x_0) be a pointed polyhedron, let $\mathbf{C}/X$ be a Lie quasi crossed module with additive feedback over (X, x_0), let (α, β) be a connection-curvature pair for $\mathbf{C}/X$, and let (σ, τ) be a kite in (X, x_0). Let us write*

$$g := \text{MI}\big(\alpha \mid \partial(\sigma, \tau)\big) \in G$$

and

$$h := \text{MI}(\alpha, \beta \mid \sigma, \tau) \in H.$$

Then

$$\Psi(g) = \text{Ad}_H(h)$$

as automorphisms of the pointed manifold $(H, 1)$. In particular, $\Psi(g)$ is a group automorphism of H.

Proof. According to Theorem 6.2.1 we have $g = \Phi_0(h)$, and by condition $(*)$ of Definition 5.1.1 we know that

$$\Psi(g) = \Psi(\Phi_0(h)) = \text{Ad}_H(h).$$

$\square$

Corollary 6.3.2. *In the situation of Theorem 6.3.1, suppose (σ', τ') is another kite in (X, x_0). We get a closed string $\partial(\sigma', \tau')$ based at x_0, and a kite $(\partial(\sigma', \tau') * \sigma, \tau)$.*

Then, writing

$$h' := \text{MI}(\alpha, \beta \mid \sigma', \tau') \in H,$$

one has

$$\text{MI}\big(\alpha, \beta \mid \partial(\sigma', \tau') * \sigma, \tau\big) = \text{Ad}_H(h')(h)$$

in H.

See Figure 26.

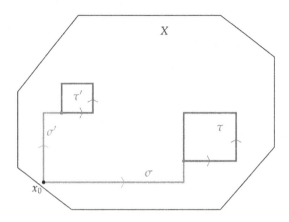

FIGURE 26. Illustration for Corollary 6.3.2.

Proof. Let $\rho := \partial(\sigma', \tau')$, which is a closed string based at x_0, and let $g' := \mathrm{MI}(\alpha \,|\, \rho) \in G$. By Theorem 6.3.1 we have $\Psi(g') = \mathrm{Ad}_H(h')$, so this is a group automorphism of H. Consider the "moving of the base point" corresponding to ρ. Since $\Psi(g')$ is a group automorphism of H, it follows that $H^{g'} = H$, and

$$\mathrm{MI}^{g'}(\alpha, \beta \,|\, \sigma, \tau) = \mathrm{MI}(\alpha, \beta \,|\, \sigma, \tau) = h.$$

But from Theorem 5.4.8 we get

$$\Psi(g')\big(\mathrm{MI}^{g'}(\alpha, \beta \,|\, \sigma, \tau)\big) = \mathrm{MI}(\alpha, \beta \,|\, \rho * \sigma, \tau).$$

$\square$

7 Square Puzzles

In this chapter we work with the pointed polyhedron $(X, x_0) := (\mathbf{I}^2, v_0)$. Let us fix some Lie quasi crossed module with additive feedback

$$\mathbf{C}/\mathbf{I}^2 = (G, H, \Psi, \Phi_0, \Phi_X)$$

over $(\mathbf{I}^2, v_0)$, and a piecewise smooth connection-curvature pair (α, β) in $\mathbf{C}/\mathbf{I}^2$. See Definitions 5.2.2 and 5.3.1. We will show that under certain homotopical restrictions, moving little square kites around inside $\mathbf{I}^2$ does not alter the multiplicative surface integral.

7.1 The Free Monoid with Involution

It will be helpful for us to have some terminology for abstract words and cancellation.

Recall that a *monoid* is a unital semigroup. Suppose M, N are monoids. By homomorphism of monoids we mean a function $\phi : M \to N$ that preserves the multiplications and the units.

Let S be some set, possibly infinite, whose elements we consider as symbols. A *word* in S is by definition a finite sequence $w = (s_1, \ldots, s_n)$ of elements of S. Thus w is a function

$$w : \{1, \ldots, n\} \to S.$$

The natural number n is the *length* of w. We denote the set of all these words by $\mathrm{Wrd}(S)$. This is a monoid under the operation of concatenation, which we denote by $*$, and the unit is the empty word $1 := ()$. We consider S as a subset of $\mathrm{Wrd}(S)$, namely the words of length 1. In fact $\mathrm{Wrd}(S)$ a free monoid: any function $S \to M$, where M is a monoid, extends uniquely to a homomorphism of monoids $\mathrm{Wrd}(S) \to M$.

Next suppose $s = (s_1, \ldots, s_n)$ is a sequence of distinct elements of S. We refer to such a sequence of distinct elements as an *alphabet*. We denote by

$$\mathrm{Wrd}(s) = \mathrm{Wrd}(s_1, \ldots, s_n)$$

the subset of $\mathrm{Wrd}(S)$ consisting of the words in the alphabet s only. More precisely, a word w of length m belongs to $\mathrm{Wrd}(s)$ if and only if the function $w : \{1, \ldots, m\} \to S$ factors as $w = s \circ \tilde{w}$, for a function (necessarily unique)

$$\tilde{w} : \{1,\ldots,m\} \to \{1,\ldots,n\}.$$

The elements of $\mathrm{Wrd}(s)$ are denoted by $w(s)$, and are called *words in s*. Clearly $\mathrm{Wrd}(s)$ is a sub-monoid of $\mathrm{Wrd}(S)$; and it is also a free monoid, on the n symbols $s_1,\ldots,s_n$. Given a homomorphism of monoids $\phi : \mathrm{Wrd}(S) \to M$, and a word $w(s) \in \mathrm{Wrd}(s)$, we shall use the following "substitution notation":

$$w(\gamma) = w(\gamma_1,\ldots,\gamma_n) := \phi(w(s)) \in M,$$

where γ is the sequence

$$\gamma = (\gamma_1,\ldots,\gamma_n) := (\phi(s_1),\ldots,\phi(s_n))$$

in M.

Let M be a monoid. By an anti-automorphism of M we mean a bijection $\psi : M \to M$ that reverses the order of multiplication, and preserves the unit. An *involution* of the monoid M is an anti-automorphism ψ such that $\psi \circ \psi = \mathrm{id}$. The pair (M,ψ) is then called a *monoid with involution*.

Let S be a set. Let S^{-1} be a new copy of S; i.e. S^{-1} is a set disjoint from S, equipped with a bijection $\psi_S : S \xrightarrow{\simeq} S^{-1}$. This bijection extends to an involution ψ_S of the set $S \cup S^{-1}$. We define

$$\mathrm{Wrd}^{\pm 1}(S) := \mathrm{Wrd}(S \cup S^{-1}).$$

This monoid has a canonical involution ψ_S, extending the involution on the set $S \cup S^{-1}$. We sometimes write $w^{-1} := \psi_S(w)$ for a word $w \in \mathrm{Wrd}^{\pm 1}(S)$. Note however that the product $w * w^{-1}$ is not 1 (unless $w = 1$, the empty word). Indeed, the length of $w * w^{-1}$ is twice the length of w (as words in $S \cup S^{-1}$).

The monoid $\mathrm{Wrd}^{\pm 1}(S)$ is a *free monoid with involution*. Here is what this means: let (M,ψ) be any monoid with involution, and let $f : S \to M$ be a function. Then there is a unique homomorphism of monoids $\phi : \mathrm{Wrd}^{\pm 1}(S) \to M$ that commutes with the involutions and extends f.

Given a sequence $s = (s_1,\ldots,s_n)$ of distinct elements of S, we write

$$\mathrm{Wrd}^{\pm 1}(s) = \mathrm{Wrd}^{\pm 1}(s_1,\ldots,s_n)$$
$$:= \mathrm{Wrd}(s_1,\ldots,s_n,s_1^{-1},\ldots,s_n^{-1}) \subset \mathrm{Wrd}^{\pm 1}(S).$$

This is also a free monoid with involution.

Suppose the words w and w' in $\mathrm{Wrd}^{\pm 1}(S)$ satisfy this condition:

$$w = v_1 * u * v_2 \quad \text{and} \quad w' = v_1 * v_2,$$

where v_1 and v_2 are some elements of $\mathrm{Wrd}^{\pm 1}(S)$, and u is either $s * s^{-1}$ or $s^{-1} * s$ for some $s \in S$. Then we say that w' is *obtained from w by cancellation*. The equivalence relation generated by this condition is called *cancellation equivalence*, and is denoted by $\sim_{\mathrm{can}}$. Thus words $w, w' \in \mathrm{Wrd}(S)$ satisfy $w \sim_{\mathrm{can}} w'$ if and only if there are words

$$w = w_0, w_1, \ldots, w_r = w'$$

such that for each i either w_{i+1} is obtained from w_i by cancellation, or vice versa. Note that the set of equivalence classes $\mathrm{Wrd}^{\pm 1}(S) / \sim_{\mathrm{can}}$ is a free group, with basis the image of S.

When we talk about $\mathrm{Wrd}^{\pm 1}(s_1, \ldots, s_n)$, we always mean implicitly that it is the free monoid with involution on some sequence $(s_1, \ldots, s_n)$ of distinct elements in some set S (possibly $S = \{s_1, \ldots, s_n\}$).

If Γ is a group, then by default we put on it the involution $\psi(\gamma) := \gamma^{-1}$. In this way Γ becomes a monoid with involution.

Let (M, ψ) be a monoid with involution. Given an element $m \in M$, define the operation $\mathrm{Ad}(m) : M \to M$ by

$$(7.1.1) \qquad \mathrm{Ad}(m)(m') := m * m' * \psi(m).$$

Warning: $\mathrm{Ad}(m)$ is not a homomorphism of monoids (unless M is a group).

Let Y be a finite graph (i.e. a finite 1-dimensional cellular complex), with base point y_0 that's a vertex, and n edges $\tau_1, \ldots, \tau_n$. Choose an orientation on each τ_i (i.e. a homeomorphism $\mathbf{I}^1 \cong \tau_i$). The reversely oriented cell is denoted by τ_i^{-1}. Take an alphabet $s = (s_1, \ldots, s_n)$. Any word $w(s) \in \mathrm{Wrd}^{\pm 1}(s)$ gives rise, by the evaluation $s_i \mapsto \tau_i$ and $s_i^{-1} \mapsto \tau_i^{-1}$, a sequence of oriented cells $w(\tau_1, \ldots, \tau_n)$, which might or might not be a path in Y.

Lemma 7.1.2. *In the situation described above, suppose $w(\tau_1, \ldots, \tau_n)$ is a closed path in Y based at y_0, such that $[w(\tau_1, \ldots, \tau_n)] = 1$ in the fundamental group $\pi_1(Y, y_0)$. Then $w(s) \sim_{\mathrm{can}} 1$ in $\mathrm{Wrd}^{\pm 1}(s)$.*

Proof. We learned this proof from Y. Glasner. First consider a pointed tree $(\tilde{Y}, \tilde{y}_0)$, and a sequence $(\tilde{\rho}_1, \ldots, \tilde{\rho}_l)$ of edges in $\tilde{Y}$ which is a path starting at $\tilde{y}_0$. If this path is closed, then it is cancellation equivalent to a point. This can be seen by induction on l. Indeed, let i be an index such that the endpoint of the path $(\tilde{\rho}_1, \ldots, \tilde{\rho}_i)$ is at maximal distance from the base vertex $\tilde{y}_0$. Then we must have $\tilde{\rho}_{i+1} = \tilde{\rho}_i^{-1}$; so we can cancel these two edges, yielding a shorter closed path.

Getting back to our problem, write $(\rho_1, \ldots, \rho_l) := [w(\tau_1, \ldots, \tau_n)]$, where $\rho_i \in \{\tau_1^{\pm 1}, \ldots, \tau_n^{\pm 1}\}$. Then $w(s) \sim_{\mathrm{can}} 1$ if and only if $(\rho_1, \ldots, \rho_l)$ is cancellation equivalent to a point. Let $p : (\tilde{Y}, \tilde{y}_0) \to (Y, y_0)$ be the universal covering map; so $(\tilde{Y}, \tilde{y}_0)$ is a pointed tree. Since the closed path $(\rho_1, \ldots, \rho_l)$ is trivial in the fundamental group of Y, it lifts to a closed path $(\tilde{\rho}_1, \ldots, \tilde{\rho}_l)$ based at $\tilde{y}_0$ in the tree $\tilde{Y}$. By the first paragraph the path $(\tilde{\rho}_1, \ldots, \tilde{\rho}_l)$ is cancellation equivalent to a point. Hence so is $(\rho_1, \ldots, \rho_l)$. $\qquad\square$

7.2 Generating Sequences

Recall that for $k \geq 0$, the k-th binary subdivision $\mathrm{sd}^k \, \mathbf{I}^2$ of $\mathbf{I}^2$ is the cellular decomposition of $\mathbf{I}^2$ into 4^k squares, each of side $(\frac{1}{2})^k$. The 1-skeleton of $\mathrm{sd}^k \, \mathbf{I}^2$ is the topological space $\mathrm{sk}_1 \, \mathrm{sd}^k \, \mathbf{I}^2$, and its fundamental group, based at v_0, is $\pi_1(\mathrm{sk}_1 \, \mathrm{sd}^k \, \mathbf{I}^2)$. For a closed string σ based at v_0 and patterned on $\mathrm{sd}^k \, \mathbf{I}^2$ we denote by $[\sigma]$ the corresponding element of $\pi_1(\mathrm{sk}_1 \, \mathrm{sd}^k \, \mathbf{I}^2)$.

Definition 7.2.1. Let $k \in \mathbb{N}$, and let
$$\rho^\natural = ((\sigma_1^\natural, \tau_1^\natural), \ldots, (\sigma_{4^k}^\natural, \tau_{4^k}^\natural))$$
be a sequence of kites in $(\mathbf{I}^2, v_0)$, all patterned on $\mathrm{sd}^k \, \mathbf{I}^2$. For any i let
$$a_i^\natural := [\partial(\sigma_i^\natural, \tau_i^\natural)] \in \pi_1(\mathrm{sk}_1 \, \mathrm{sd}^k \, \mathbf{I}^2).$$

(1) If the sequence $(a_1^\natural, \ldots, a_{4^k}^\natural)$ is a basis of the group $\pi_1(\mathrm{sk}_1 \, \mathrm{sd}^k \, \mathbf{I}^2)$, then we say that $\rho^\natural$ is a *generating sequence for* $\mathrm{sd}^k \, \mathbf{I}^2$.

(2) Let (σ, τ) be a kite in $(\mathbf{I}^2, v_0)$ patterned on $\mathrm{sd}^k \, \mathbf{I}^2$. If $\tau = \tau_i^\natural$ for some i then we say that (σ, τ) is *aligned with* $\rho^\natural$.

The k-th binary tessellation $\mathrm{tes}^k \, \mathbf{I}^2$ (see Definition 4.2.5) is an example of a generating sequence for $\mathrm{sd}^k \, \mathbf{I}^2$.

Lemma 7.2.2. *Let $\rho^\natural$ be a generating sequence for $\mathrm{sd}^k \, \mathbf{I}^2$, and let $a_i^\natural =$ as in the definition above. Let $w(s) \in \mathrm{Wrd}^{\pm 1}(s) = \mathrm{Wrd}^{\pm 1}(s_1, \ldots, s_{4^k})$ be a word such that*
$$w(a_1^\natural, \ldots, a_{4^k}^\natural) = 1$$
in the group $\pi_1(\mathrm{sk}_1 \, \mathrm{sd}^k \, \mathbf{I}^2)$. Then
$$w(s) \sim_{\mathrm{can}} 1$$
in $\mathrm{Wrd}^{\pm 1}(s)$.

Proof. This is because the sequence $(a_1^\natural, \ldots, a_{4^k}^\natural)$ is a basis of the free group $\pi_1(\mathrm{sk}_1 \, \mathrm{sd}^k \, \mathbf{I}^2)$. $\qquad\square$

Lemma 7.2.3. *Let σ_1, σ_2 be linear strings in $\mathbf{I}^2$, and let $\tau : \mathbf{I}^2 \to \mathbf{I}^2$ be a linear map, all patterned on $\mathrm{sd}^k \mathbf{I}^2$ for some $k \geq 0$.*

(1) *Suppose σ_1, σ_2 are closed strings based at v_0, and $[\sigma_1] = [\sigma_2]$ in $\pi_1(\mathrm{sk}_1 \mathrm{sd}^k \mathbf{I}^2)$. Then*

$$\mathrm{MI}(\alpha \,|\, \sigma_1) = \mathrm{MI}(\alpha \,|\, \sigma_1)$$

in G.

(2) *Suppose that $\sigma_i(v_0) = v_0$ and $\sigma_i(v_1) = \tau(v_0)$ for $i = 1, 2$; so that (σ_i, τ) are kites in $\mathbf{I}^2$ and $\sigma_2 * \sigma_1^{-1}$ is a closed string based at v_0. If $[\sigma_2 * \sigma_1^{-1}] = 1$ in $\pi_1(\mathrm{sk}_1 \mathrm{sd}^k \mathbf{I}^2)$, then*

$$\mathrm{MI}(\alpha, \beta \,|\, \sigma_1, \tau) = \mathrm{MI}(\alpha, \beta \,|\, \sigma_2, \tau)$$

in H.

Proof. (1) Take $n := 2^{k+1} \cdot (2^k + 1)$, the number of 1-cells in $\mathrm{sd}^k \mathbf{I}^2$. Let us denote these 1-cells by $\tau_1, \ldots, \tau_n$, and let's choose an orientation for these cells, as in Lemma 7.1.2. There are unique words

$$w_1(s), w_2(s) \in \mathrm{Wrd}^{\pm 1}(s) = \mathrm{Wrd}^{\pm 1}(s_1, \ldots, s_n)$$

such that $w_i(\tau_1, \ldots, \tau_n) = \sigma_i$ as strings. Lemma 7.1.2 implies that $w_1(s) \sim_{\mathrm{can}} w_2(s)$ in $\mathrm{Wrd}^{\pm 1}(s)$. Now Definition 3.5.7 and Proposition 3.5.8 tell us that

$$\mathrm{MI}\big(\alpha \,|\, w_1(\tau_1, \ldots, \tau_n)\big) = \mathrm{MI}\big(\alpha \,|\, w_2(\tau_1, \ldots, \tau_n)\big).$$

(2) Let $g_i := \mathrm{MI}(\alpha \,|\, \sigma_i) \in G$. By part (1) we know that

$$g_2 \cdot g_1^{-1} = \mathrm{MI}(\alpha \,|\, \sigma_2) \cdot \mathrm{MI}(\alpha \,|\, \sigma_1^{-1}) = \mathrm{MI}(\alpha \,|\, \sigma_2 * \sigma_1^{-1}) = 1;$$

so that $g_1 = g_2$. Let $g := g_1 = g_2$.

Let σ_0 denote the empty string. So $\sigma_i * \sigma_0 = \sigma_i$ for $i = 1, 2$. Now according to Theorem 5.4.8 we have

$$\mathrm{MI}(\alpha, \beta \,|\, \sigma_i, \tau) = \Psi(g)\big(\mathrm{MI}^g(\alpha, \beta \,|\, \sigma_0, \tau)\big)$$

for $i = 1, 2$. But the right hand side is independent of i. $\qquad\qquad \square$

Lemma 7.2.4. *Let $\rho^\natural$ be a generating sequence for $\mathrm{sd}^k \mathbf{I}^2$, let $a_i^\natural$ be as in Definition 7.2.1, and let*

$$h_i^\natural := \mathrm{MI}(\alpha, \beta \,|\, \sigma_i^\natural, \tau_i^\natural) \in H.$$

Suppose (σ, τ) is a kite in $(\mathbf{I}^2, v_0)$ patterned on $\mathrm{sd}^k \mathbf{I}^2$ and aligned with $\rho^\natural$. Then there is a word

$$w(s) \in \mathrm{Wrd}^{\pm 1}(s) = \mathrm{Wrd}^{\pm 1}(s_1, \ldots, s_{4^k})$$

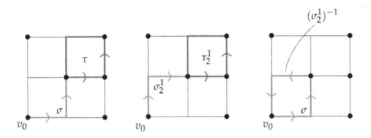

FIGURE 27. A kite (σ, τ) patterned on $\mathrm{sd}^1 \, \mathrm{I}^2$ and aligned with $\mathrm{tes}^1 \, \mathrm{I}^2$, the corresponding kite (σ_i^1, τ_i^1), and the closed string $\sigma * (\sigma_i^1)^{-1}$. Here $i = 2$.

such that

$$[\partial(\sigma, \tau)] = w(a_1^\natural, \ldots, a_{4k}^\natural)$$

in $\pi_1(\mathrm{sk}_1 \, \mathrm{sd}^k \, \mathrm{I}^2)$, and

$$\mathrm{MI}(\alpha, \beta \mid \sigma, \tau) = w(h_1^\natural, \ldots, h_{4k}^\natural)$$

in H.

Proof. Let i be an index such that $\tau = \tau_i^\natural$. Then $\sigma * (\sigma_i^\natural)^{-1}$ is a closed string based at v_0 and patterned on $\mathrm{sd}^k \, \mathrm{I}^2$. (See Figure 27 for an illustration where $k = 1 \, \rho^\natural = \mathrm{tes}^1 \, \mathrm{I}^2$.) There is a word $u(s) \in \mathrm{Wrd}^{\pm 1}(s)$ such that

$$[\sigma * (\sigma_i^\natural)^{-1}] = u(a_1^\natural, \ldots, a_{4k}^\natural)$$

in the group $\pi_1(\mathrm{sk}_1 \, \mathrm{sd}^k \, \mathrm{I}^2)$.

We can also consider the monoid M of finite sequences of closed strings patterned on $\mathrm{sd}^k \, \mathrm{I}^2$ and based at v_0, where composition is concatenation of sequences, and the involution is given by reversal of order and (3.5.4). Consider the sequence

$$\partial \rho^\natural := \big(\partial(v_1^\natural, \tau_1^\natural), \ldots, \partial(\sigma_{4k}^\natural, \tau_{4k}^\natural)\big).$$

The evaluation $u(\partial \rho^\natural)$ of the word $u(s)$ on the sequence $\partial \rho^\natural$ is an element of M. And we have

(7.2.5) $$[u(\partial \rho^\natural)] = u(a_1^\natural, \ldots, a_{4k}^\natural)$$

in the group $\pi_1(\mathrm{sk}_1 \, \mathrm{sd}^k \, \mathrm{I}^2)$.

Take

$$w(s) := u(s) * s_i * u(s)^{-1} \in \mathrm{Wrd}^{\pm 1}(s).$$

It is easy to check that

$$[\partial(\sigma,\tau)] = w(a_1^\natural, \ldots, a_{4k}^\natural)$$

holds in $\pi_1(\mathrm{sk}_1 \, \mathrm{sd}^k \, \mathbf{I}^2)$. Next, using Corollary 6.3.2 recursively, equation (7.2.5) and Lemma 7.2.3(2), we obtain

$$w(h_1^\natural, \ldots, h_{4k}^\natural) = u(h_1^\natural, \ldots, h_{4k}^\natural) \cdot h_i^\natural \cdot u(h_1^\natural, \ldots, h_{4k}^\natural)^{-1}$$
$$= \mathrm{MI}(\alpha, \beta \mid u(\partial \rho^\natural) * \sigma_i^\natural, \tau_i^\natural)$$
$$= \mathrm{MI}(\alpha, \beta \mid \sigma, \tau).$$

$\square$

Lemma 7.2.6. *Let $\rho^\natural$ be a generating sequence for $\mathrm{sd}^k \, \mathbf{I}^2$, and let*

$$\rho = ((\sigma_1, \tau_1), \ldots, (\sigma_m, \tau_m))$$

be some sequence of kites in $(\mathbf{I}^2, v_0)$ patterned on $\mathrm{sd}^k \, \mathbf{I}^2$ and aligned with $\rho^\natural$. Write

$$a_i := [\partial(\sigma_i, \tau_i)] \in \pi_1(\mathrm{sk}_1 \, \mathrm{sd}^k \, \mathbf{I}^2)$$

and

$$h_i := \mathrm{MI}(\alpha, \beta \mid \sigma_i, \tau_i) \in H.$$

Let

$$u(s) \in \mathrm{Wrd}^{\pm 1}(s) = \mathrm{Wrd}^{\pm 1}(s_1, \ldots, s_m)$$

be a word such that

$$u(a_1, \ldots, a_m) = 1.$$

Then

$$u(h_1, \ldots, h_m) = 1.$$

Proof. Using Lemma 7.2.4, for each $i \in \{1, \ldots, m\}$ we can find a word

$$w_i(t) \in \mathrm{Wrd}^{\pm 1}(t) = \mathrm{Wrd}^{\pm 1}(t_1, \ldots, t_{4k})$$

such that

$$a_i = w_i(a_1^\natural, \ldots, a_{4k}^\natural)$$

and

$$h_i = w_i(h_1^\natural, \ldots, h_{4k}^\natural)$$

(with notation as in the previous lemmas). Define

$$w(t) := u(w_1(t), \ldots, w_m(t)) \in \mathrm{Wrd}^{\pm 1}(t).$$

Then

$$w(a_1^\natural, \ldots, a_{4k}^\natural) = 1$$

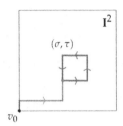 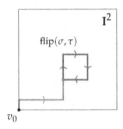

FIGURE 28. A kite (σ, τ) in $(\mathbf{I}^2, v_0)$, and the kite $\mathrm{flip}(\sigma, \tau)$.

in $\pi_1(\mathrm{sk}_1 \mathrm{sd}^k \mathbf{I}^2)$, and

$$w(h_1^\natural, \ldots, h_{4k}^\natural) = u(h_1, \ldots, h_m)$$

in H.

Now according to Lemma 7.2.2 we have $w(t) \sim_{\mathrm{can}} 1$ in $\mathrm{Wrd}^{\pm 1}(t)$. Hence by cancellation in the group H we get $w(h_1^\natural, \ldots, h_{4k}^\natural) = 1$. □

7.3 The Flip

The flip of $\mathbf{I}^2$ is the linear automorphism whose action on vertices is

$$\mathrm{flip}(v_0, v_1, v_2) := (v_0, v_2, v_1).$$

Given a kite (σ, τ) in $(\mathbf{I}^2, v_0)$, its flip is the kite

$$\mathrm{flip}(\sigma, \tau) := (\sigma, \tau \circ \mathrm{flip}).$$

See Figure 28 for an illustration.

Note that

$$\mathrm{flip}(\sigma, \tau) = (\sigma, \tau) \circ \mathrm{flip}(\sigma_1^0, \tau_1^0),$$

where (σ_1^0, τ_1^0) is the basic kite. By associativity of kite composition it follows that given any two kites (σ_1, τ_1) and (σ_2, τ_3) in $\mathbf{I}^2$, one has

(7.3.1) $$\mathrm{flip}\big((\sigma_1, \tau_1) \circ (\sigma_2, \tau_2)\big) = (\sigma_1, \tau_1) \circ \mathrm{flip}(\sigma_2, \tau_2).$$

If (σ, τ) is a kite patterned on $\mathrm{sd}^k \mathbf{I}^2$ for some k, then the effect of flipping is:

(7.3.2) $$[\partial \mathrm{flip}(\sigma, \tau)] = [\partial(\sigma, \tau)]^{-1}$$

in $\pi_1(\mathrm{sk}_1 \mathrm{sd}^k \mathbf{I}^2)$.

Lemma 7.3.3. *Let (σ, τ) be a kite in $(\mathbf{I}^2, v_0)$, and let $k \geq 0$. Then*

$$(7.3.4) \qquad \mathrm{MI}(\alpha, \beta \mid \mathrm{flip}(\sigma, \tau)) = \prod_{i=4^k}^{1} \mathrm{MI}(\alpha, \beta \mid \mathrm{flip}((\sigma, \tau) \circ (\sigma_i^k, \tau_i^k))).$$

Note that the order of the product is reversed!

Proof. By moving the base point from v_0 to $x_0 := \tau(v_0)$ along the string σ, and using Theorem 5.4.8, we now have to prove that (7.3.4) holds for a kite (σ, τ) in the pointed polyhedron $(\mathbf{I}^2, x_0)$, and moreover σ is the empty string. Next we use Proposition 4.5.3 for the map of pointed polyhedra $\tau : (\mathbf{I}^2, v_0) \rightarrow (\mathbf{I}^2, x_0)$ to reduce to proving (7.3.4) for $(\sigma, \tau) = (\sigma_1^0, \tau_1^0)$, the basic kite in $(\mathbf{I}^2, v_0)$.

So we have to prove that

$$(7.3.5) \qquad \mathrm{MI}(\alpha, \beta \mid \mathrm{flip}(\sigma_1^0, \tau_1^0)) = \prod_{i=4^k}^{1} \mathrm{MI}(\alpha, \beta \mid \mathrm{flip}(\sigma_i^k, \tau_i^k))$$

holds.

For any $i \in \{1, \ldots, 4^k\}$ define

$$(\sigma_i^\natural, \tau_i^\natural) := \mathrm{flip}(\sigma_1^0, \tau_1^0) \circ (\sigma_i^k, \tau_i^k).$$

Then

$$\rho^\natural := ((\sigma_1^\natural, \tau_1^\natural), \ldots, (\sigma_{4^k}^\natural, \tau_{4^k}^\natural))$$

is a generating sequence for $\mathrm{sd}^k \mathbf{I}^2$. By Proposition 4.5.1 we know that

$$(7.3.6) \qquad \mathrm{MI}(\alpha, \beta \mid \mathrm{flip}(\sigma_1^0, \tau_1^0)) = \prod_{i=1}^{4^k} h_i^\natural,$$

where $h_i^\natural$ are as in Lemma 7.2.4. And it is clear from Definition 4.2.2(2) and formula (7.3.2) that

$$(7.3.7) \qquad \prod_{i=1}^{4^k} a_i^\natural = [\partial \mathbf{I}^2]^{-1},$$

where $a_i^\natural$ are as in Definition 7.2.1.

Consider another sequence of kites

$$\rho := ((\sigma_1, \tau_1), \ldots, (\sigma_{4^k}, \tau_{4^k})),$$

where we define

$$(\sigma_i, \tau_i) := \mathrm{flip}(\sigma_i^k, \tau_i^k).$$

It is not hard to show, by induction on k, that these kites are aligned with the generating sequence $\rho^{\natural}$. Let a_i and h_i be as in Lemma 7.2.6, where we take $m := 4^k$ of course. Again by induction on k one shows that

$$\prod_{i=4^k}^{1} a_i = [\partial \mathbf{I}^2]^{-1}.$$

Combining this with (7.3.7) we deduce

$$a_{4^k} \cdots a_2 \cdot a_1 \cdot (a_{4^k}^{\natural})^{-1} \cdots (a_2^{\natural})^{-1} \cdot (a_1^{\natural})^{-1} = 1$$

in $\pi_1(\mathrm{sk}_1 \, \mathrm{sd}^k \, \mathbf{I}^2)$. Now Lemma 7.2.6 says that

$$h_{4^k} \cdots h_2 \cdot h_1 \cdot (h_{4^k}^{\natural})^{-1} \cdots (h_2^{\natural})^{-1} \cdot (h_1^{\natural})^{-1} = 1$$

in H. Using (7.3.6) we get

$$\prod_{i=4^k}^{1} h_i = \mathrm{MI}(\alpha, \beta \,|\, \mathrm{flip}(\sigma_1^0, \tau_1^0)).$$

This is precisely (7.3.5). □

Definition 7.3.8. In this chapter, a kite (σ, τ) in $(\mathbf{I}^2, v_0)$ will be called (α, β)-*tiny* if it is a square kite, $\mathrm{len}(\sigma) \leq 5$, and

$$\mathrm{side}(\tau) < \min\big(\epsilon_4(\alpha, \beta), \, \epsilon_5(\alpha, \Psi_\natural)\big).$$

See Lemma 6.1.3 and Proposition 3.5.13 regarding the constants $\epsilon_4(\alpha, \beta)$ and $\epsilon_5(\alpha, \Psi_\natural)$.

Lemma 7.3.9. *Let (σ, τ) be a kite in $(\mathbf{I}^2, v_0)$. Then*

$$(7.3.10) \qquad \mathrm{MI}(\alpha, \beta \,|\, \mathrm{flip}(\sigma, \tau)) = \mathrm{MI}(\alpha, \beta \,|\, \sigma, \tau)^{-1}.$$

Proof. Step 1. Assume (σ, τ) is (α, β)-tiny and $\beta|_{\tau(\mathbf{I}^2)}$ is smooth. We may assume that $\epsilon := \mathrm{side}(\tau)$ is positive (since the case $\epsilon = 0$ is trivial).

Since the flip reverses the orientation, it follows that

$$\mathrm{RP}_0(\alpha, \beta \,|\, \mathrm{flip}(\sigma, \tau)) = \mathrm{RP}_0(\alpha, \beta \,|\, \sigma, \tau)^{-1};$$

cf. Definition 4.3.3. Next, by applying Proposition 4.5.2 to both (σ, τ) and $\mathrm{flip}(\sigma, \tau)$, we have
(7.3.11)
$$\big\| \log_H\big(\mathrm{MI}(\alpha, \beta \,|\, \mathrm{flip}(\sigma, \tau))\big) - \log_H\big(\mathrm{MI}(\alpha, \beta \,|\, \sigma, \tau)^{-1}\big) \big\| \leq 2 \cdot c_2(\alpha, \beta) \cdot \epsilon^4.$$

Step 2. Again we assume that (σ, τ) is (α, β)-tiny and $\epsilon := \mathrm{side}(\tau)$ is positive; but we do not assume smoothness.

Take $k \geq 0$. For each index $i \in \{1, \ldots, 4^k\}$ we let

$$(\sigma_i, \tau_i) := (\sigma, \tau) \circ (\sigma_i^k, \tau_i^k) = \mathrm{tes}_i^k(\sigma, \tau).$$

Note that $\mathrm{side}(\tau_i) = (\frac{1}{2})^k \cdot \epsilon$. The subsets $\mathrm{good}(\tau, k)$ and $\mathrm{bad}(\tau, k)$ of $\{1, \ldots, 4^k\}$ were defined in Definition 4.4.12. According to Lemma 4.4.13, there are constants

$$a_j := a_j(\alpha, \beta, \tau(\mathbf{I}^2))$$

such that

$$|\mathrm{bad}(\tau, k)| \leq a_0 + a_1 \cdot 2^k.$$

If i is a good index, then by step 1 we know that

$$\left\| \log_H \left(\mathrm{MI}(\alpha, \beta \mid \mathrm{flip}(\sigma_i, \tau_i)) \right) - \log_H \left(\mathrm{MI}(\alpha, \beta \mid \sigma_i, \tau_i)^{-1} \right) \right\|$$
$$\leq 2 \cdot c_2(\alpha, \beta) \cdot (\tfrac{1}{2})^{4k} \cdot \epsilon^4.$$

And if i is a bad index, then by Proposition 4.5.2(1) we have

(7.3.12)
$$\left\| \log_H \left(\mathrm{MI}(\alpha, \beta \mid \mathrm{flip}(\sigma_i, \tau_i)) \right) - \log_H \left(\mathrm{MI}(\alpha, \beta \mid \sigma_i, \tau_i)^{-1} \right) \right\|$$
$$\leq 2 \cdot c_1(\alpha, \beta) \cdot (\tfrac{1}{2})^{2k} \cdot \epsilon^2.$$

The conditions in property (ii) of Theorem 2.1.2 are satisfied. Hence, using Lemma 7.3.3 and Proposition 4.5.1, we obtain these estimates:

$$\left\| \log_H \left(\mathrm{MI}(\alpha, \beta \mid \mathrm{flip}(\sigma, \tau)) \right) - \log_H \left(\mathrm{MI}(\alpha, \beta \mid \sigma, \tau)^{-1} \right) \right\|$$

$$= \left\| \log_H \left(\prod_{i=4^k}^{1} \mathrm{MI}(\alpha, \beta \mid \mathrm{flip}(\sigma_i, \tau_i)) \right) \right.$$

$$\left. - \log_H \left(\prod_{i=4^k}^{1} \mathrm{MI}(\alpha, \beta \mid \sigma_i, \tau_i)^{-1} \right) \right\|$$

$$\leq |\mathrm{good}(\tau, k)| \cdot 2 \cdot c_2(\alpha, \beta) \cdot (\tfrac{1}{2})^{4k} \cdot \epsilon^4$$
$$+ |\mathrm{bad}(\tau, k)| \cdot 2 \cdot c_1(\alpha, \beta) \cdot (\tfrac{1}{2})^{2k} \cdot \epsilon^2$$
$$\leq 4^k \cdot 2 \cdot c_2(\alpha, \beta) \cdot (\tfrac{1}{2})^{4k} \cdot \epsilon^4 + (a_0 + a_1 \cdot 2^k) \cdot 2 \cdot c_1(\alpha, \beta) \cdot (\tfrac{1}{2})^{2k} \cdot \epsilon^2.$$

As $k \to \infty$ the last term goes to 0. Hence (7.3.10) holds in this case.

Step 3. Now (σ, τ) is an arbitrary kite in $(\mathbf{I}^2, v_0)$. We may assume that $\tau(\mathbf{I}^2)$ is 2-dimensional, for otherwise things are trivial. As done in the beginning of the proof of Lemma 7.3.3, we can modify the setup so that (σ, τ) is a square kite and $\mathrm{len}(\sigma) \leq 1$. Take k large enough so that all the kites

$$(\sigma_i, \tau_i) := (\sigma, \tau) \circ (\sigma_i^k, \tau_i^k),$$

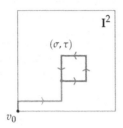 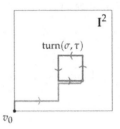

FIGURE 29. A kite (σ, τ) in $(\mathbf{I}^2, v_0)$, and the kite $\mathrm{turn}(\sigma, \tau)$.

$i \in \{1, \ldots, 4^k\}$, are (α, β)-tiny. By Step 2 we know that

$$\mathrm{MI}\big(\alpha, \beta \,|\, \mathrm{flip}(\sigma_i, \tau_i)\big) = \mathrm{MI}(\alpha, \beta \,|\, \sigma_i, \tau_i)^{-1}$$

holds for all i. Hence, using Lemma 7.3.3 and Proposition 4.5.1, we conclude that

$$\mathrm{MI}\big(\alpha, \beta \,|\, \mathrm{flip}(\sigma, \tau)\big) = \prod_{i=4^k}^{1} \mathrm{MI}\big(\alpha, \beta \,|\, \mathrm{flip}(\sigma_i, \tau_i)\big)$$

$$= \prod_{i=4^k}^{1} \mathrm{MI}(\alpha, \beta \,|\, \sigma_i, \tau_i)^{-1} = \left(\prod_{i=1}^{4^k} \mathrm{MI}(\alpha, \beta \,|\, \sigma_i, \tau_i) \right)^{-1}$$

$$= \mathrm{MI}(\alpha, \beta \,|\, \sigma, \tau)^{-1}.$$

$\square$

7.4   The Turn

The (counterclockwise) turn of $\mathbf{I}^2$ is the linear automorphism $\mathrm{turn} : \mathbf{I}^2 \to \mathbf{I}^2$ defined on vertices by

$$\mathrm{turn}(v_0, v_1, v_2) := (v_1, (1,1), v_0).$$

The turn of the basic kite (σ_1^0, τ_1^0) is

$$\mathrm{turn}(\sigma_1^0, \tau_1^0) := \big((v_0, v_1), \mathrm{turn}\big).$$

Given any kite (σ, τ) in $(\mathbf{I}^2, v_0)$, its turn is the kite

$$\mathrm{turn}(\sigma, \tau) := (\sigma, \tau) \circ \mathrm{turn}(\sigma_1^0, \tau_1^0).$$

See Figure 29.

Observe that for kites (σ_1, τ_1) and (σ_2, τ_2) in $(\mathbf{I}^2, v_0)$, one has

(7.4.1) $\mathrm{turn}\big((\sigma_1, \tau_1) \circ (\sigma_2, \tau_2)\big) = (\sigma_1, \tau_1) \circ \mathrm{turn}(\sigma_2, \tau_2).$

If (σ, τ) is patterned on $\mathrm{sd}^k \, \mathbf{I}^2$, then the boundaries satisfy

(7.4.2) $$[\partial \, \mathrm{turn}(\sigma, \tau)] = [(\sigma, \tau)]$$

in $\pi_1(\mathrm{sk}_1 \, \mathrm{sd}^k \, \mathbf{I}^2)$.

Lemma 7.4.3. *Let (σ, τ) be a kite in $(\mathbf{I}^2, v_0)$, and let $k \geq 0$. Then*

$$\mathrm{MI}(\alpha, \beta \mid \mathrm{turn}(\sigma, \tau)) = \prod_{i=1}^{4^k} \mathrm{MI}(\alpha, \beta \mid \mathrm{turn}((\sigma, \tau) \circ (\sigma_i^k, \tau_i^k))).$$

Proof. The proof is very similar to that of Lemma 7.3.3. As we showed there, it suffices to consider the case $(\sigma, \tau) = (\sigma_1^0, \tau_1^0)$. So we have to prove

(7.4.4) $$\mathrm{MI}(\alpha, \beta \mid \mathrm{turn}(\sigma_1^0, \tau_1^0)) = \prod_{i=1}^{4^k} \mathrm{MI}(\alpha, \beta \mid \mathrm{turn}(\sigma_i^k, \tau_i^k)).$$

For any $i \in \{1, \ldots, 4^k\}$ let

$$(\sigma_i^\natural, \tau_i^\natural) := \mathrm{turn}(\sigma_1^0, \tau_1^0) \circ (\sigma_i^k, \tau_i^k).$$

The sequence

$$\rho^\natural := ((\sigma_1^\natural, \tau_1^\natural), \ldots, (\sigma_{4^k}^\natural, \tau_{4^k}^\natural))$$

is a generating sequence of kites for $\mathrm{sd}^k \, \mathbf{I}^2$. Define elements $a_i^\natural$ and $h_i^\natural$ as in Definition 7.2.1 and Lemma 7.2.4, with respect to the this new generating sequence $\rho^\natural$. By Proposition 4.5.1 we know that

(7.4.5) $$\mathrm{MI}(\alpha, \beta \mid \mathrm{turn}(\sigma_1^0, \tau_1^0)) = \prod_{i=1}^{4^k} h_i^\natural.$$

And by induction on k it is not hard to see that

(7.4.6) $$\prod_{i=1}^{4^k} a_i^\natural = [\partial \mathbf{I}^2].$$

Next consider the sequence of kites

$$\rho := ((\sigma_1, \tau_1), \ldots, (\sigma_{4^k}, \tau_{4^k})),$$

where

$$(\sigma_i, \tau_i) := \mathrm{turn}(\sigma_i^k, \tau_i^k).$$

These kites are all aligned with $\rho^\natural$. Take a_i and h_i defined as in Lemma 7.2.6. We then have

$$\prod_{i=1}^{4^k} a_i = [\partial \mathbf{I}^2].$$

Applying Lemma 7.2.6 we conclude that

$$\prod_{i=1}^{4^k} h_i = \prod_{i=1}^{4^k} h_i^{\natural}.$$

Combining this equation with (7.4.5) we can deduce that equation (7.4.4) is true.   □

Lemma 7.4.7. Let (σ, τ) be a kite in $(\mathbf{I}^2, v_0)$. Then

$$(7.4.8) \qquad \mathrm{MI}\big(\alpha, \beta \,|\, \mathrm{turn}(\sigma, \tau)\big) = \mathrm{MI}(\alpha, \beta \,|\, \sigma, \tau).$$

Proof. The proof is organized like that of Lemma 7.3.9, so we allow ourselves to be less detailed.

Step 1. Here we assume that (σ, τ) is an (α, β)-tiny kite, the number $\epsilon := \mathrm{side}(\tau)$ is positive, and $\beta|_{\tau(\mathbf{I}^2)}$ is smooth. Going back to Definition 4.3.3, and using its notation mixed with the present notation, we have

$$\mathrm{RP}_0(\alpha, \beta \,|\, \sigma, \tau) = \exp_H\big(\epsilon^2 \cdot \Psi_\natural(g)(\tilde{\beta}(z))\big).$$

Recall that

$$g = \mathrm{MI}\big(\alpha \,|\, \sigma * (\tau \circ \sigma_{\mathrm{pr}})\big).$$

The turn does not change the area of the square $Z = \tau(\mathbf{I}^2)$, nor its midpoint z; it only changes the string that leads from v_0 to z. Indeed, the string that is related to $\mathrm{turn}(\sigma, \tau)$ is $\sigma * (\tau \circ \rho')$, where ρ' is the string

$$\rho' := (v_0, v_1) * (v_1, y) * (y, w),$$

in $\mathbf{I}^2$ with $y := (\frac{1}{2}, 1)$ and $w := (\frac{1}{2}, \frac{1}{2})$. The formula for the Riemann product is

$$\mathrm{RP}_0(\alpha, \beta \,|\, \mathrm{turn}(\sigma, \tau)) = \exp_H\big(\epsilon^2 \cdot \Psi_\natural(g')(\tilde{\beta}(z))\big),$$

with

$$g' := \mathrm{MI}\big(\alpha \,|\, \sigma * (\tau \circ \rho')\big).$$

Let $\rho'' := \sigma_{\mathrm{pr}}^{-1} * \rho'$, which is a closed string based at w. The group element

$$g'' := \mathrm{MI}\big(\alpha \,|\, \tau \circ \rho''\big)$$

satisfies $g' = g \cdot g''$. Since $\mathrm{len}(\tau \circ \rho'') = 3\epsilon$, we know by Proposition 3.5.13 that

$$\big\| \Psi_\natural(g'') - 1 \big\| \le c_5(\alpha, \Psi_\natural) \cdot 3\epsilon.$$

But by Proposition 3.5.10 we have

$$\big\| \Psi_\natural(g) \big\| \le \exp\big(c_4(\alpha, \Psi_\natural) \cdot 5\big).$$

Therefore

$$\| \log_H(\mathrm{RP}_0(\alpha, \beta \mid \mathrm{turn}(\sigma, \tau))) - \log_H(\mathrm{RP}_0(\alpha, \beta \mid \sigma, \tau)) \|$$
$$= \epsilon^2 \cdot (\Psi_\mathfrak{h}(g) \circ (\Psi_\mathfrak{h}(g'') - 1)) (\tilde{\beta}(z))$$
$$\leq \epsilon^3 \cdot c,$$

where we write

$$c := \exp(c_4(\alpha, \Psi_\mathfrak{h}) \cdot 5) \cdot 3 \cdot c_5(\alpha, \Psi_\mathfrak{h}) \cdot \|\beta\|_{\mathrm{Sob}}.$$

Combining this estimate with Proposition 4.5.2(2), we obtain

(7.4.9)
$$\| \log_H(\mathrm{MI}(\alpha, \beta \mid \mathrm{turn}(\sigma, \tau))) - \log_H(\mathrm{MI}(\alpha, \beta \mid \sigma, \tau)) \|$$
$$\leq 2 \cdot c_2(\alpha, \beta) \cdot \epsilon^4 + c \cdot \epsilon^3.$$

Step 2. In this step we assume that (σ, τ) is an (α, β)-tiny kite and $\epsilon :=$ side(τ) is positive; but no smoothness is assumed.

Take $k \geq 0$, and define kites (σ_i, τ_i) and sets good(τ, k) and bad(τ, k) like in step 2 of the proof of Lemma 7.3.9. If $i \in$ good(τ, k) then by equation (7.4.9) in step 1 we know that

$$\| \log_H(\mathrm{MI}(\alpha, \beta \mid \mathrm{turn}(\sigma_i, \tau_i))) - \log_H(\mathrm{MI}(\alpha, \beta \mid \sigma_i, \tau_i)) \|$$
$$\leq 2 \cdot c_2(\alpha, \beta) \cdot (\tfrac{1}{2})^{4k} \cdot \epsilon^4 + c \cdot (\tfrac{1}{2})^{3k} \cdot \epsilon^3.$$

For $i \in$ bad(τ, k) we use the estimate (7.3.12). As in the proof of Lemma 7.3.9, but using Lemma 7.4.3 instead of Lemma 7.3.3, we arrive at the estimate

$$\| \log_H(\mathrm{MI}(\alpha, \beta \mid \mathrm{turn}(\sigma, \tau))) - \log_H(\mathrm{MI}(\alpha, \beta \mid \sigma, \tau)) \|$$
$$= \left\| \log_H\left(\prod_{i=1}^{4^k} \mathrm{MI}(\alpha, \beta \mid \mathrm{turn}(\sigma_i, \tau_i)) \right) - \log_H\left(\prod_{i=1}^{4^k} \mathrm{MI}(\alpha, \beta \mid \sigma_i, \tau_i) \right) \right\|$$
$$\leq |\mathrm{good}(\tau, k)| \cdot \left(2 \cdot c_2(\alpha, \beta) \cdot (\tfrac{1}{2})^{4k} \cdot \epsilon^4 + c \cdot (\tfrac{1}{2})^{3k} \cdot \epsilon^3 \right)$$
$$+ |\mathrm{bad}(\tau, k)| \cdot \left(2 \cdot c_1(\alpha, \beta) \cdot (\tfrac{1}{2})^{2k} \cdot \epsilon^2 \right)$$
$$\leq 4^k \cdot \left(2 \cdot c_2(\alpha, \beta) \cdot (\tfrac{1}{2})^{4k} \cdot \epsilon^4 + c \cdot (\tfrac{1}{2})^{3k} \cdot \epsilon^3 \right)$$
$$+ (a_0 + a_1 \cdot 2^k) \cdot \left(2 \cdot c_1(\alpha, \beta) \cdot (\tfrac{1}{2})^{2k} \cdot \epsilon^2 \right).$$

As $k \to \infty$ the last term goes to 0. Hence (7.4.8) holds in this case.

Step 3. For an arbitrary kite (σ, τ) in $(\mathbf{I}^2, v_0)$ we prove that (7.4.8) holds using step 2, as was done in step 3 of the proof of Lemma 7.3.9, but using Lemma 7.4.3 instead of Lemma 7.3.3. $\qquad \square$

7.5 Putting it all Together

Theorem 7.5.1. *Let*
$$\mathbf{C}/\mathbf{I}^2 = (G, H, \Psi, \Phi_0, \Phi_X)$$
be a Lie quasi crossed module over $(X, x_0) := (\mathbf{I}^2, v_0)$, *let* (α, β) *be a connection-curvature pair in* $\mathbf{C}/\mathbf{I}^2$, *and let*
$$\rho = ((\sigma_1, \tau_1), \ldots, (\sigma_m, \tau_m))$$
be a sequence of kites in $(\mathbf{I}^2, v_0)$ *patterned on* $\mathrm{sd}^k \mathbf{I}^2$, *for some* $k \in \mathbb{N}$. *For* $i \in \{1, \ldots, m\}$ *define*
$$a_i := [\partial(\sigma_i, \tau_i)] \in \pi_1(\mathrm{sk}_1 \mathrm{sd}^k \mathbf{I}^2)$$
and
$$h_i := \mathrm{MI}(\alpha, \beta \mid \sigma_i, \tau_i) \in H.$$

Suppose
$$w(s) \in \mathrm{Wrd}^{\pm 1}(s) = \mathrm{Wrd}^{\pm 1}(s_1, \ldots, s_m)$$
is a word such that
$$w(a_1, \ldots, a_m) = 1$$
in $\pi_1(\mathrm{sk}_1 \mathrm{sd}^k \mathbf{I}^2)$. *Then*
$$w(h_1, \ldots, h_m) = 1$$
in H.

Proof. We can find $l_i \in \{0, 1\}$ and $j_i \in \{0, 1, 2, 3\}$, such that the kites
$$(\sigma_i', \tau_i') := \mathrm{flip}^{l_i}\left(\mathrm{turn}^{j_i}(\sigma_i, \tau_i)\right)$$
are aligned with the generating sequence $\mathrm{tes}^k \mathbf{I}^2$. Here the exponents l_i and j_i refer to iteration of the corresponding operation. Let $e_i := (-1)^{l_i}$,
$$a_i' := [\partial(\sigma_i', \tau_i')] \in \pi_1(\mathrm{sk}_1 \mathrm{sd}^k \mathbf{I}^2)$$
and
$$h_i' := \mathrm{MI}(\alpha, \beta \mid \sigma_i', \tau_i') \in H.$$
Then $a_i^{e_i} = a_i'$, and by Lemmas 7.3.9 and 7.4.7 we also have $h_i^{e_i} = h_i'$.

Consider the word
$$u(s) := w(s_1^{e_1}, \ldots, s_m^{e_m}) \in \mathrm{Wrd}^{\pm 1}(s).$$
Then
$$w(a_1, \ldots, a_m) = u(a_1', \ldots, a_m')$$
and
$$w(h_1, \ldots, h_m) = u(h_1', \ldots, h_m').$$
Now we can finish the proof with the use of Lemma 7.2.6. $\square$

Corollary 7.5.2. *Let*

$$\mathbf{C}/\mathbf{I}^2 = (G, H, \Psi, \Phi_0, \Phi_X)$$

be a Lie quasi crossed module over $(X, x_0) := (\mathbf{I}^2, v_0)$, *let* (α, β) *be a connection-curvature pair in* $\mathbf{C}/\mathbf{I}^2$, *and let*

$$\rho = ((\sigma_1, \tau_1), \ldots, (\sigma_{4^k}, \tau_{4^k}))$$

be a tessellation of $\mathbf{I}^2$ *patterned on* $\mathrm{sd}^k \, \mathbf{I}^2$ *(cf. Definition 4.2.2). Then*

$$\prod_{i=1}^{4^k} \mathrm{MI}(\alpha, \beta \mid \sigma_i, \tau_i) = \mathrm{MI}(\alpha, \beta \mid \mathbf{I}^2).$$

Proof. This is an immediate consequence of the theorem, taking $m := 4^k + 1$,

$$w(s) := \left(\prod_{i=1}^{m-1} s_i \right) \cdot s_m^{-1}$$

and (σ_m, τ_m) the basic kite. $\qquad\qquad\qquad\qquad\qquad\qquad\qquad\qquad \square$

8 Stokes Theorem in Dimension Three

The goal of this chapter is to prove Theorem 8.6.6, which is the first version of the main result of the book. (The second version, dealing with the triangular case, is Theorem 9.3.5.)

8.1 Balloons and their Boundaries

Definition 8.1.1. Let (X, x_0) be a pointed polyhedron. A *linear quadrangular balloon* in (X, x_0) is a pair (σ, τ), where σ is a string in X (see Definition 3.5.1), and $\tau : \mathbf{I}^3 \to X$ is a linear map. The conditions are that $\sigma(v_0) = x_0$ and $\sigma(v_1) = \tau(v_0)$.

In other words, a linear quadrangular balloon is the 3-dimensional version of a linear quadrangular kite. See Figure 30 for an illustration.

All balloons in this chapter are linear quadrangular ones; so we shall simply call them balloons. (This will change in Chapter 9.) If the image of τ is a cube in X, then we call (σ, τ) a *cubical balloon*. The length of the side of τ is denoted by $\mathrm{side}(\tau)$. If $\mathrm{side}(\tau) > 0$ then (σ, τ) is said to be *nondegenerate*.

If (σ, τ) is a balloon in (X, x_0), and (σ', τ') is a balloon (resp. a kite) in $(\mathbf{I}^3, v_0)$, then the composition $(\sigma, \tau) \circ (\sigma', \tau')$ (defined like (4.1.2)) is a balloon (resp. a kite) in (X, x_0).

The k-th binary subdivision of $\mathbf{I}^3$ is its cellular decomposition into 8^k little cubes, each of side $(\frac{1}{2})^k$. We denote this decomposition by $\mathrm{sd}^k \mathbf{I}^3$.

A map $\sigma : \mathbf{I}^p \to \mathbf{I}^3$ is said to be *patterned on* $\mathrm{sd}^k \mathbf{I}^3$ if it is linear, and its image is a p-cell in $\mathrm{sd}^k \mathbf{I}^3$. A string $\sigma = (\sigma_1, \ldots, \sigma_n)$ in $(\mathbf{I}^3, v_0)$ is said to be patterned on $\mathrm{sd}^k \mathbf{I}^3$ if for every i the map $\sigma_i : \mathbf{I}^1 \to \mathbf{I}^3$ is patterned on $\mathrm{sd}^k \mathbf{I}^3$. A kite or balloon (σ, τ) in $(\mathbf{I}^3, v_0)$ is said to be patterned on $\mathrm{sd}^k \mathbf{I}^3$ if the string σ and the map $\tau : \mathbf{I}^p \to \mathbf{I}^3$ ($p = 2, 3$) are patterned on $\mathrm{sd}^k \mathbf{I}^3$.

The k-th binary tessellation of $\mathbf{I}^3$ is the sequence

$$(8.1.2) \qquad \mathrm{tes}^k \mathbf{I}^3 = \left(\mathrm{tes}_1^k \mathbf{I}^3, \ldots, \mathrm{tes}_{8^k}^k \mathbf{I}^3\right) = \left((\sigma_1^k, \tau_1^k), \ldots, (\sigma_{8^k}^k, \tau_{8^k}^k)\right)$$

of balloons patterned on $\mathrm{sd}^k \mathbf{I}^3$, defined as follows. For $k = 0$ we have the basic balloon (σ_1^0, τ_1^0), where σ_1^0 is the empty string, and $\tau_1^0 : \mathbf{I}^3 \to \mathbf{I}^3$ is the identity map.

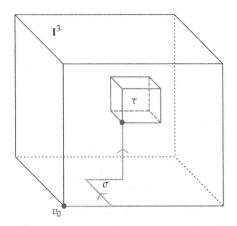

FIGURE 30. A cubical balloon (σ, τ) in the pointed polyhedron $(\mathbf{I}^3, v_0)$.

For $k = 1$ we choose, once and for all, a sequence

$$\mathrm{tes}^1 \, \mathbf{I}^3 = \big((\sigma_1^1, \tau_1^1), \ldots, (\sigma_8^1, \tau_8^1)\big)$$

balloons patterned on $\mathrm{sd}^1 \, \mathbf{I}^3$, satisfying these conditions:

(a) Each of the 3-cells of $\mathrm{sd}^1 \, \mathbf{I}^3$ occurs exactly once as $\tau_i^1(\mathbf{I}^3)$ for some i.

(b) The maps τ_i^1 are positively oriented.

(c) The length of each string σ_i^1 is at most 3.

This can be done of course.

For $k \geq 1$ we use the recursive definition

$$\mathrm{tes}^{k+1} \, \mathbf{I}^3 := (\mathrm{tes}^1 \, \mathbf{I}^3) \circ (\mathrm{tes}^k \, \mathbf{I}^3).$$

Here we use the convention (3.1.1) for composition of sequences.

Given a balloon (σ, τ) in a pointed polyhedron (X, x_0), an numbers $k \in \mathbb{N}$, $i \in \{1, \ldots, 8^k\}$, let

(8.1.3) $\mathrm{tes}_i^k(\sigma, \tau) := (\sigma, \tau) \circ \mathrm{tes}_i^k \, \mathbf{I}^3 = (\sigma, \tau) \circ (\sigma_i^k, \tau_i^k).$

The k-th binary tessellation of (σ, τ) is the sequence (of length 8^k) of balloons

(8.1.4) $\mathrm{tes}^k(\sigma, \tau) := (\sigma, \tau) \circ (\mathrm{tes}^k \, \mathbf{I}^3) = \big(\mathrm{tes}_1^k(\sigma, \tau), \ldots, \mathrm{tes}_{8^k}^k(\sigma, \tau)\big).$

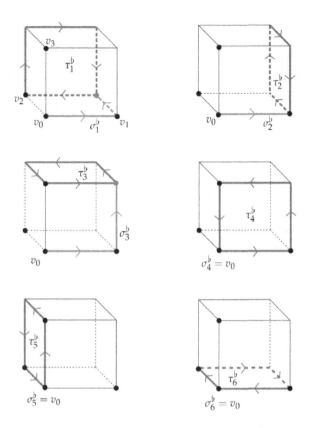

FIGURE 31. The kites $(\sigma_i^\flat, \tau_i^\flat)$, $i = 1, \ldots, 6$, that make up the boundary of $\mathbf{I}^3$.

Definition 8.1.5.   (1) For $i \in \{1, \ldots, 6\}$ let $\partial_i \mathbf{I}^3 := (\sigma_i^\flat, \tau_i^\flat)$ be the kites depicted in Figure 31. The *boundary of* $\mathbf{I}^3$ is the sequence of kites

$$\partial \mathbf{I}^3 = (\partial_1 \mathbf{I}^3, \ldots, \partial_6 \mathbf{I}^3) = ((\sigma_1^\flat, \tau_1^\flat), \ldots, (\sigma_6^\flat, \tau_6^\flat)).$$

(2) Given a balloon (σ, τ) in a pointed polyhedron (X, x_0), let

$$\partial_i(\sigma, \tau) := (\sigma, \tau) \circ \partial_i \mathbf{I}^3 = (\sigma, \tau) \circ (\sigma_i^\flat, \tau_i^\flat).$$

The *boundary* of (σ, τ) is the sequence of kites

$$\partial(\sigma, \tau) := (\partial_1(\sigma, \tau), \ldots, \partial_6(\sigma, \tau))$$

in (X, x_0).

If (σ, τ) is a balloon in $(\mathbf{I}^3, v_0)$ patterned on $\mathrm{sd}^k \mathbf{I}^3$, then $\partial(\sigma, \tau)$ is a sequence (of length 6) of kites patterned on $\mathrm{sd}^k \mathbf{I}^3$, and $\partial\partial(\sigma, \tau)$ is a sequence of closed strings patterned on $\mathrm{sd}^k \mathbf{I}^3$.

Definition 8.1.6. Let (X, x_0) be a pointed polyhedron, let $\mathbf{C}/X$ be a Lie quasi crossed module with additive feedback over (X, x_0), and let (α, β) be a piecewise smooth connection-curvature pair in $\mathbf{C}/X$. Given a balloon (σ, τ) in (X, x_0), we define

$$\mathrm{MI}(\alpha, \beta \mid \partial(\sigma, \tau)) := \prod_{1=1}^{6} \mathrm{MI}(\alpha, \beta \mid \partial_i(\sigma, \tau)) \in H.$$

8.2 Some Algebraic Topology

Let us denote by $\pi_1(\mathrm{sk}_1 \, \mathrm{sd}^k \mathbf{I}^3)$ the fundamental group of the topological space $\mathrm{sk}_1 \, \mathrm{sd}^k \mathbf{I}^3$, based at v_0. If σ is a closed string patterned on $\mathrm{sd}^k \mathbf{I}^3$ and based at v_0, then we denote its class in $\pi_1(\mathrm{sk}_1 \, \mathrm{sd}^k \mathbf{I}^3)$ by $[\sigma]$.

We now look at homology groups. Fix $k \geq 0$. On each p-cell τ in $\mathrm{sd}^k \mathbf{I}^3$, $p \in \{0, 1, 2\}$, let us choose an orientation (there are two options for $p > 0$). We denote by $C_p(\mathrm{sk}_2 \, \mathrm{sd}^k \mathbf{I}^3)$ the free abelian group based on these oriented cells, called the group of p-chains. The direct sum on all p is a complex (with the usual boundary operator), and the p-th homology is the singular homology $H_p(\mathrm{sk}_2 \, \mathrm{sd}^k \mathbf{I}^3)$ of the topological space $\mathrm{sk}_2 \, \mathrm{sd}^k \mathbf{I}^3$. The homology class of a cycle $\sum_i n_i \tau_i$ is denoted by $[\sum_i n_i \tau_i]$. Observe that a string $\sigma = (\sigma_1, \ldots, \sigma_n)$ patterned on $\mathrm{sd}^k \mathbf{I}^3$ represents an element $\sum_i \sigma_i \in C_1(\mathrm{sk}_2 \, \mathrm{sd}^k \mathbf{I}^3)$. And a linear map $\tau : \mathbf{I}^2 \to \mathbf{I}^3$ patterned on $\mathrm{sd}^k \mathbf{I}^3$ represents an element $\tau \in C_2(\mathrm{sk}_2 \, \mathrm{sd}^k \mathbf{I}^3)$.

Recall the free monoid with involution $\mathrm{Wrd}^{\pm 1}(s)$ from Section 7.1.

Lemma 8.2.1. Let $\big((\sigma_1, \tau_1), \ldots, (\sigma_n, \tau_n)\big)$ be a sequence of kites patterned on $\mathrm{sk}_2 \, \mathrm{sd}^k \mathbf{I}^3$, for some $k \geq 0$. Write

$$a_i := [\partial(\sigma_i, \tau_i)] \in \pi_1(\mathrm{sk}_1 \, \mathrm{sd}^k \mathbf{I}^3).$$

Let $w(s) \in \mathrm{Wrd}^{\pm 1}(s_1, \ldots, s_n)$ be a word such that $w(a_1, \ldots, a_n) = 1$. Then the 2-chain

$$w(\tau_1, \ldots, \tau_n) \in C_2(\mathrm{sk}_2 \, \mathrm{sd}^k \mathbf{I}^3)$$

is a cycle.

Proof. Say m is the number of 1-cells in $\mathrm{sd}^k \mathbf{I}^3$. We choose orientations, and arrange these oriented 1-cells in a sequence $(\rho_1, \ldots, \rho_m)$. For every i let $u_i(t) \in \mathrm{Wrd}^{\pm 1}(t_1, \ldots, t_m)$ be the word such that

$$\partial(\sigma_i, \tau_i) = u_i(\rho_1, \ldots, \rho_m)$$

as strings patterned on $\mathrm{sd}^k \mathbf{I}^3$. Define

$$u(t) := w(u_1(t), \ldots, u_n(t)) \in \mathrm{Wrd}^{\pm 1}(t).$$

Then $u(\rho_1, \ldots, \rho_m)$ is a closed string, and its homotopy class is trivial in the group $\pi_1(\mathrm{sk}_1 \mathrm{sd}^k \mathbf{I}^3)$. According to Lemma 7.1.2 we have $u(t) \sim_{\mathrm{can}} 1$ in $\mathrm{Wrd}^{\pm 1}(t)$.

On the other hand we have $\partial(\tau_i) = \partial(\sigma_i, \tau_i)$ as additive 1-chains, i.e. as elements of $C_1(\mathrm{sk}_2 \mathrm{sd}^k \mathbf{I}^3)$. Hence

$$\partial(w(\tau_1, \ldots, \tau_n)) = w(\partial(\sigma_1, \tau_1), \ldots, \partial(\sigma_n, \tau_n)) = u(\rho_1, \ldots, \rho_m) = 0$$

as elements of $C_2(\mathrm{sk}_2 \mathrm{sd}^k \mathbf{I}^3)$.   □

Lemma 8.2.2. *Recall the boundary of $\mathbf{I}^3$ from Definition 8.1.5. For any i let*

$$a_i^\flat := [\partial(\sigma_i^\flat, \tau_i^\flat)] \in \pi_1(\mathrm{sk}_1 \mathrm{sd}^0 \mathbf{I}^3)$$

and

$$b_i^\flat := \tau_i^\flat \in C_2(\mathrm{sk}_1 \mathrm{sd}^0 \mathbf{I}^3).$$

Then:

(1) *The fundamental group $\pi_1(\mathrm{sk}_1 \mathrm{sd}^0 \mathbf{I}^3)$ is generated by the sequence of elements $(a_1^\flat, \ldots, a_6^\flat)$, and there is one relation:*

$$\prod_{i=1,\ldots,6} a_i^\flat = 1.$$

Thus $\pi_1(\mathrm{sk}_1 \mathrm{sd}^0 \mathbf{I}^3)$ is a free group of rank 5.

(2) *The homology group $\mathrm{H}_2(\mathrm{sk}_2 \mathrm{sd}^0 \mathbf{I}^3)$ is a free abelian group of rank 1, with basis $[\sum_{i=1,\ldots,6} \tau_i]$.*

Proof. This is obvious from looking at the pictures.   □

Given a balloon (σ, τ), its boundary $\partial(\sigma, \tau)$ is a sequence of 6 kites, and $\partial\partial(\sigma, \tau)$ is a sequence of 6 closed strings.

Lemma 8.2.3.

(1) *The fundamental group $\pi_1(\mathrm{sk}_1 \mathrm{sd}^1 \mathbf{I}^3)$ is generated by the sequence of closed strings*

$$\partial\partial(\mathrm{tes}^1 \mathbf{I}^3) = (\partial\partial(\sigma_1^k, \tau_1^k), \ldots, \partial\partial(\sigma_{8k}^k, \tau_{8k}^k)).$$

This sequence has length 48. There are 20 relations, and they are of two kinds:

(a) For any two distinct kites $(\sigma_i, \tau_i), (\sigma_j, \tau_j)$ in $\partial\partial(\mathrm{tes}^1\, \mathbf{I}^3)$ such that $\tau_i(\mathbf{I}^2) = \tau_j(\mathbf{I}^2)$, namely for any of the 12 interior faces of $\mathrm{sd}^1\, \mathbf{I}^3$, there is a relation

$$[\partial(\sigma_i, \tau_i)] = [g_i] \cdot [\partial(\sigma_j, \tau_j)^{-1}] \cdot [g_i^{-1}]$$

for some word g_i in the 48 generators.

(b) For any of the 8 balloons (σ_i^k, τ_i^k) in $\mathrm{tes}^1\, \mathbf{I}^3$ there is a relation as in Lemma 8.2.2(1).

Thus $\pi_1(\mathrm{sk}_1\, \mathrm{sd}^1\, \mathbf{I}^3)$ is a free group of rank 28.

(2) The homology group $H_2(\mathrm{sk}_2\, \mathrm{sd}^1\, \mathbf{I}^3)$ is a free abelian group of rank 8, with basis the 8 boundaries of the 8 balloons in $\mathrm{tes}^1\, \mathbf{I}^3$.

Proof. Use Lemma 8.2.2 together with the Van-Campen and Mayer-Vietoris theorems.　　　　　　　　　　　　　　　　　　　　　　□

8.3　Inert Forms

Let (X, x_0) be a pointed polyhedron, and let

$$\mathbf{C}/X = (G, H, \Psi, \Phi_0, \Phi_X)$$

be a Lie quasi crossed module with additive feedback over (X, x_0). Recall that the additive feedback Φ_X is an element of $\mathcal{O}_{\mathrm{pws}}(X) \otimes \mathrm{Hom}(\mathfrak{h}, \mathfrak{g})$, so for any point $x \in X$ we have a linear function $\Phi_X(x) : \mathfrak{h} \to \mathfrak{g}$.

Suppose $Z \subset X$ is some sub-polyhedron. As in equation (4.6.4), but restricting to Z, for every p we have an $\mathcal{O}_{\mathrm{pws}}(Z)$-linear homomorphism

$$\Phi_X|_Z : \Omega^p_{\mathrm{pws}}(Z) \otimes \mathfrak{h} \to \Omega^p_{\mathrm{pws}}(Z) \otimes \mathfrak{g}.$$

Note that for $p = 0$, an element

$$f \in \Omega^0_{\mathrm{pws}}(Z) \otimes \mathfrak{h} = \mathcal{O}_{\mathrm{pws}}(Z) \otimes \mathfrak{h}$$

is a piecewise smooth function $f : Z \to \mathfrak{h}$; and in this case

$$\Phi_X|_Z(f)(z) = \Phi_X(z)(f(z)) \in \mathfrak{g}$$

for any point $z \in Z$.

Definition 8.3.1. Let $Z \subset X$ be a subpolyhedron.

(1) A function $f \in \mathcal{O}_{\mathrm{pws}}(Z) \otimes \mathfrak{h}$ is called an *inert function* (relative to $\mathbf{C}/X$) if $\Phi_X|_Z(f) = 0$.

(2) A form $\gamma \in \Omega^p_{\mathrm{pws}}(Z) \otimes \mathfrak{h}$ is called an *inert p-form* (relative to $\mathbf{C}/X$) if $\Phi_X|_Z(\gamma) = 0$.

Let $\gamma \in \Omega^p_{\mathrm{pws}}(X) \otimes \mathfrak{h}$. Choose some linear coordinate system $s = (s_1, \ldots, s_n)$ on X. Also choose a smoothing triangulation $\{X_j\}_{j \in J}$ for γ; so that $\gamma|_{X_j} \in \Omega^p(X_j) \otimes \mathfrak{h}$ for every j. See Section 1.6 for details. For each index j let $f_{j,i} \in \mathcal{O}(X_j) \otimes \mathfrak{h}$ be the coefficients of $\gamma|_{X_j}$ relative to s, in the sense of Definition 1.3.4. Namely

(8.3.2) $$\gamma|_{X_j} = \sum_i f_{j,i} \cdot \mathrm{d}s_{i_1} \wedge \cdots \wedge \mathrm{d}s_{i_p} \in \Omega^p(X_j) \otimes \mathfrak{h},$$

where $i = (i_1, \ldots, i_p)$ runs over the set of strictly increasing multi-indices in $\{1, \ldots, n\}^p$.

Lemma 8.3.3. *In the situation above, the following are equivalent:*

(i) *The form $\gamma \in \Omega^p_{\mathrm{pws}}(X) \otimes \mathfrak{h}$ is inert.*

(ii) *The functions $f_{j,i} \in \mathcal{O}(X_j) \otimes \mathfrak{h}$ are all inert.*

We omit the easy proof.

Definition 8.3.4. The closed subgroup

$$H_0 := \mathrm{Ker}(\Phi_0) \subset H$$

is called the *inertia subgroup* (relative to $\mathbf{C}/X$). Its Lie algebra

$$\mathfrak{h}_0 := \mathrm{Lie}(H_0) = \mathrm{Ker}(\mathrm{Lie}(\Phi_0))$$

is called the *inertia subalgebra*.

Proposition 8.3.5. *The subgroup H_0 is central in H, and the subalgebra $\mathfrak{h}_0$ is central in $\mathfrak{h}$.*

Proof. Take $h \in H_0$, so $\Phi_0(h) = 1$. By the Pfeiffer condition (i.e. condition $(*)$ of Definition 5.1.1) we have

$$\mathrm{Ad}_H(h) = \Psi(\Phi_0(h)) = \Psi(1) = \mathrm{id}_H,$$

which says that

$$h \in \mathrm{Ker}(\mathrm{Ad}_H) = Z(H).$$

Since $H_0 \subset Z(H)$ it follows that

$$\mathfrak{h}_0 \subset \mathrm{Lie}(Z(H)) \subset Z(\mathfrak{h}).$$

(Note that H could be disconnected, in which case $Z(\mathfrak{h})$ could be bigger than $\mathrm{Lie}(Z(H))$.) □

Recall the notion of tame connection (Definition 5.3.4).

Lemma 8.3.6. *Let α be a tame connection for $\mathbf{C}/X$, let σ be a closed string in X based at x_0, and let $g := \mathrm{MI}(\alpha \,|\, \sigma) \in G$. Then for any $\lambda \in \mathfrak{h}_0$ one has*

$$\Psi_{\mathfrak{h}}(g)(\lambda) = \lambda.$$

Proof. Let m be the number of pieces in σ, and choose k large enough such that $m \leq 2^{k+2}$. We may append to σ a few copies of the constant map x_0 at its end, so that now σ has exactly 2^{k+2} pieces. The group element g is unchanged.

Denote by σ' the closed string in $\mathbf{I}^2$ which is "k-th subdivision" of the boundary $\partial \mathbf{I}^2$, based at v_0; so σ' has $2^{k+2} = 4 \cdot 2^k$ pieces. We can construct a piecewise linear map $f : \mathbf{I}^2 \to X$ such that $f(v_0) = x_0$, and $f \circ \sigma' = \rho$ as strings. According to Propositions 3.5.9 and 3.3.22(3) we have

$$g = \mathrm{MI}(f^*(\alpha) \,|\, \sigma') = \mathrm{MI}(f^*(\alpha) \,|\, \partial \mathbf{I}^2).$$

Let $\beta \in \Omega^2_{\mathrm{pws}}(X) \otimes \mathfrak{h}$ be such that (α, β) is a connection-curvature pair for $\mathbf{C}/X$. By Proposition 5.3.3 the pair $(f^*(\alpha), f^*(\beta))$ is a connection-curvature pair in $f^*(\mathbf{C}/X)$. Hence by Theorem 6.2.1 we have $g = \Phi_0(h)$, where

$$h := \mathrm{MI}(f^*(\alpha), f^*(\beta) \,|\, \mathbf{I}^2) \in H.$$

Since $\lambda \in \mathrm{Lie}(\mathrm{Z}(H))$ it follows that

$$\Psi_\mathfrak{h}(g)(\lambda) = \Psi_\mathfrak{h}(\Phi_0(h))(\lambda) = \mathrm{Ad}_\mathfrak{h}(h)(\lambda) = \lambda.$$

$\square$

Remark 8.3.7. The lemma above shows that the holonomy group of a tame connection α at x_0 acts trivially on the inertia subalgebra $\mathfrak{h}_0$. Hence the names.

Lemma 8.3.8. *Let $\mathbf{C}/X$ be a Lie quasi crossed module with additive feedback over (X, x_0), let α be a tame connection for $\mathbf{C}/X$, and let $Z \subset X$ be a subpolyhedron.*

(1) *Let $f \in \mathcal{O}_{\mathrm{pws}}(Z) \otimes \mathfrak{h}$ be an inert function relative to $\mathbf{C}/X$. Then there is a unique function*

$$\Psi_{\mathfrak{h},\alpha}(f) : Z \to \mathfrak{h}_0$$

such that the following condition holds:
(∗) *Let σ be a string in X, with initial point x_0 and terminal point $z := \sigma(v_1) \in Z$, and let $g := \mathrm{MI}(\alpha \,|\, \sigma) \in G$. Then*

$$\Psi_{\mathfrak{h},\alpha}(f)(z) = \Psi_\mathfrak{h}(g)(f(z)).$$

(2) *The operation $\Psi_{\mathfrak{h},\alpha}$ is linear over the ring $\mathcal{O}_{\mathrm{pws}}(Z)$.*
(3) *The function $\Psi_{\mathfrak{h},\alpha}(f)$ is continuous.*

Proof. Take a point $z \in Z$. Choose any string σ connecting x_0 to z, and let

$$\lambda := \Psi_\mathfrak{h}(g)(f(z)) \in \mathfrak{h}$$

as in condition (∗). Since α is a compatible connection (see Definition 5.2.4), and since f is inert, we know that

$$\text{Lie}(\Phi_0)(\lambda) = \Phi_X(x_0)(\lambda) = \text{Ad}_{\mathfrak{g}}(g)(\Phi_X(z)(f(z))) = 0.$$

This shows that $\lambda \in \mathfrak{h}_0$.

If we were to choose another string σ' with the same initial and terminal points, then for $g' := \text{MI}(\alpha \mid \sigma')$ we would have

$$(\Psi_{\mathfrak{h}}(g') \circ \Psi_{\mathfrak{h}}(g)^{-1})(\lambda) = \lambda,$$

this according to Lemma 8.3.6. Hence $\lambda = \Psi_{\mathfrak{h}}(g')(f(z))$. We see that λ is independent of the string σ, and we can define $\Psi_{\mathfrak{h},\alpha}(f)(z) := \lambda$. We obtain a function $\Psi_{\mathfrak{h},\alpha}(f) : Z \to \mathfrak{h}_0$. This proves part (1).

Part (2) is true because the operator $\Psi_{\mathfrak{h}}(g) : \mathfrak{h} \to \mathfrak{h}$ is linear.

It remains to prove part (3). We need to prove that the function $\Psi_{\mathfrak{h},\alpha}(f) : Z \to \mathfrak{h}_0$ is continuous. So let us fix a point $z_0 \in Z$, and a string σ starting at x_0 and ending at z_0. Let $g := \text{MI}(\alpha \mid \sigma)$. For any point $z \in Z$ let $\sigma_z : \mathbf{I}^1 \to Z$ be the unique linear map with initial point z_0 and terminal point z, and let $g_z := \text{MI}(\alpha \mid \sigma_z)$. So

$$\Psi_{\mathfrak{h},\alpha}(f)(z) = \Psi_{\mathfrak{h}}(g \cdot g_z)(f(z)) = (\Psi_{\mathfrak{h}}(g) \circ \Psi_{\mathfrak{h}}(g_z))(f(z)).$$

But according to Proposition 3.5.13 we know that $\Psi_{\mathfrak{h}}(g_z) \to \mathbf{1} \in \text{End}(\mathfrak{h})$ as $z \to z_0$; and $f(z) \to f(z_0)$ by continuity of f. Therefore

$$\Psi_{\mathfrak{h},\alpha}(f)(z) \to \Psi_{\mathfrak{h},\alpha}(f)(z_0)$$

as $z \to z_0$. $\qquad\qquad\qquad\qquad\qquad\qquad\qquad\qquad\qquad\qquad\qquad\qquad\square$

To summarize, the lemma says that there is an $\mathcal{O}_{\text{pws}}(Z)$-linear homomorphism

(8.3.9) $\Psi_{\mathfrak{h},\alpha} : \{\text{inert functions on } Z\} \to \mathcal{O}_{\text{cont}}(Z) \otimes \mathfrak{h}_0,$

where $\mathcal{O}_{\text{cont}}(Z)$ is the ring of continuous functions $Z \to \mathbb{R}$.

Recall the module $\Omega^p_{\text{pwc}}(X)$ of piecewise continuous differential forms, from Section 1.9.

Proposition 8.3.10. *Let α be a tame connection, and let $\gamma \in \Omega^p_{\text{pws}}(X) \otimes \mathfrak{h}$ be an inert form. Then there is a unique piecewise continuous form*

$$\Psi_{\mathfrak{h},\alpha}(\gamma) \in \Omega^p_{\text{pwc}}(X) \otimes \mathfrak{h}_0$$

with the following property:

(∗) *Choose a linear coordinate system* $s = (s_1, \ldots, s_n)$ *on* X, *and a smoothing triangulation* $\{X_j\}_{j \in J}$ *for* γ. *Let* $f_{j,i} \in \mathcal{O}(X_j) \otimes \mathfrak{h}$ *be the coefficients of* $\gamma|_{X_j}$, *as in* (8.3.2). *Then*

$$\Psi_{\mathfrak{h},\alpha}(\gamma)|_{X_j} = \sum_i \Psi_{\mathfrak{h},\alpha}(f_{j,i}) \cdot \mathrm{d}s_{i_1} \wedge \cdots \wedge \mathrm{d}s_{i_p}.$$

Note that by Lemma 8.3.3 the functions $f_{j,i}$ are inert, so the continuous functions $\Psi_{\mathfrak{h},\alpha}(f_{j,i}) \in \mathcal{O}_{\mathrm{cont}}(X_j)$ are defined.

Proof. This is immediate from the uniqueness of the coefficients $f_{j,i}$, and the properties of the homomorphism $\Psi_{\mathfrak{h},\alpha}$ listed in Lemma 8.3.8. $\qquad\square$

Remark 8.3.11. Actually a lot more can be said here. Presumably, one can show that

$$\mathrm{Ker}(\Phi_X) \subset \mathcal{O}_{\mathrm{pws}}(X) \otimes \mathfrak{h}$$

is the set of piecewise smooth sections of a piecewise smooth vector bundle E over X, which is a sub-bundle of the trivial vector bundle $X \times \mathfrak{h}$. And the operation $\Psi_{\mathfrak{h},\alpha}$ corresponds to a piecewise smooth isomorphism of vector bundles

$$E \xrightarrow{\cong} X \times \mathfrak{h}_0.$$

If so, then it would follow that for any inert form $\gamma \in \Omega^p_{\mathrm{pws}}(Z) \otimes \mathfrak{h}$, the form $\Psi_{\mathfrak{h},\alpha}(\gamma)$ is actually piecewise smooth (not just piecewise continuous). And we would have an $\mathcal{O}_{\mathrm{pws}}(X)$-linear bijection

$$\Psi_{\mathfrak{h},\alpha} : \{\text{inert } p\text{-forms on } Z\} \to \Omega^p_{\mathrm{pws}}(Z) \otimes \mathfrak{h}_0.$$

However we did not verify these assertions.

Suppose we are given a piecewise continuous form $\delta \in \Omega^p_{\mathrm{pwc}}(X) \otimes \mathfrak{h}$ and a piecewise linear map $\tau : I^p \to X$. Extending the formula (1.9.7) linearly to $\mathfrak{h}_0$-valued forms we obtain $\int_\tau \delta \in \mathfrak{h}_0$.

Definition 8.3.12 (Integration of Inert Forms). Let C/X be a Lie quasi crossed module with additive feedback over (X, x_0). Given a tame connection α, an inert form $\gamma \in \Omega^p_{\mathrm{pws}}(X) \otimes \mathfrak{h}$ and a piecewise linear map $\tau : I^p \to X$, we define the *twisted multiplicative integral*

$$\mathrm{MI}(\alpha, \gamma \,|\, \tau) \in H_0$$

as follows:

$$\mathrm{MI}(\alpha, \gamma \,|\, \tau) := \exp_H\left(\int_\tau \Psi_{\mathfrak{h},\alpha}(\gamma)\right).$$

Proposition 8.3.13. *Let $f : Y \to X$ and $\tau : I^p \to Y$ be piecewise linear maps between polyhedra, let $\alpha \in \Omega^1_{pws}(X) \otimes \mathfrak{g}$ be a tame connection, and let $\gamma \in \Omega^p_{pws}(X) \otimes \mathfrak{h}$ be an inert form. Then $f^*(\alpha) \in \Omega^1_{pws}(Y) \otimes \mathfrak{g}$ is a tame connection, $f^*(\gamma) \in \Omega^p_{pws}(Y) \otimes \mathfrak{h}$ is an inert form, and*

$$\mathrm{MI}(\alpha, \gamma \,|\, f \circ \tau) = \mathrm{MI}\big(f^*(\alpha), f^*(\gamma) \,|\, \tau\big).$$

Proof. The connection $f^*(\alpha)$ is tame for $f^*(\mathbf{C}/X)$ by Corollary 5.3.5. It is easy to see directly from the definitions that $f^*(\gamma)$ is inert, and moreover

$$\Psi_{\mathfrak{h}, f^*(\alpha)}(f^*(\gamma)) = f^*(\Psi_{\mathfrak{h}, \alpha}(\gamma)) \in \Omega^p_{pwc}(Y) \otimes \mathfrak{h}.$$

And integration of piecewise continuous forms commutes with pullbacks (see Section 1.9). $\qquad\square$

8.4 Combinatorics and Integration

In this section we work with the pointed polyhedron $(X, x_0) := (I^3, v_0)$. We fix a Lie quasi crossed module with additive feedback

$$\mathbf{C}/I^3 = (G, H, \Psi, \Phi_0, \Phi_X),$$

and a piecewise smooth connection-curvature pair (α, β) in $\mathbf{C}/I^3$. Let us also fix a euclidean norm $\|-\|$ on $\mathfrak{h}$, an open set $V_0(H)$ in H, and constants $\epsilon_0(H)$ and $c_0(H)$ as in Chapter 2.

Lemma 8.4.1. *Let (σ, τ) be a balloon in (I^3, v_0). Then*

$$\mathrm{MI}\big(\alpha, \beta \,|\, \partial(\sigma, \tau)\big) \in H_0.$$

Proof. Let us write $h := \mathrm{MI}\big(\alpha, \beta \,|\, \partial(\sigma, \tau)\big) \in H$ and $g := \mathrm{MI}\big(\alpha \,|\, \partial\partial(\sigma, \tau)\big) \in G$. According to Theorem 6.2.1 we have $\Phi_0(h) = g$.

On the other hand, consider the word $w(s) \in \mathrm{Wrd}^{\pm 1}(s_1, \ldots, s_{12})$ such that

$$\partial\partial I^3 = w(\rho_1, \ldots, \rho_{12}),$$

where $\rho_1, \ldots, \rho_{12}$ are the oriented 1-cells of $\mathrm{sd}^0 \, I^3$. Then $[w(\rho_1, \ldots, \rho_{12})] = 1$ in $\pi_1(\mathrm{sk}_1 \, \mathrm{sd}^0 \, I^3)$. By Lemma 7.1.2 we know that $w(s) \sim_{can} 1$ in $\mathrm{Wrd}^{\pm 1}(s)$. Now writing $g_i := \mathrm{MI}(\alpha \,|\, \rho_i)$ we have $g = w(g_1, \ldots, g_{12}) = 1$ in the group G. We see that $h \in \mathrm{Ker}(\Phi_0) = H_0$. $\qquad\square$

Lemma 8.4.2. *Take $k = 0, 1$. Let $\rho = \big((\sigma_i, \tau_i)\big)_{i=1,\ldots,n}$ be a sequence of kites in (I^3, v_0), all patterned on $\mathrm{sd}^k \, I^3$. Write*

$$a_i := [\partial(\sigma_i, \tau_i)] \in \pi_1(\mathrm{sk}_1 \, \mathrm{sd}^k \, I^3)$$

and

$$h_i := \mathrm{MI}(\alpha, \beta \,|\, \sigma_i, \tau_i) \in H.$$

Let $w(s) \in \mathrm{Wrd}^{\pm 1}(s_1, \ldots, s_n)$ be a word such that

$$w(a_1, \ldots, a_n) = 1$$

in $\pi_1(\mathrm{sk}_1 \, \mathrm{sd}^k \, \mathbf{I}^3)$ and

$$[w(\tau_1, \ldots, \tau_n)] = 0$$

in $H_2(\mathrm{sk}_2 \, \mathrm{sd}^k \, \mathbf{I}^3)$. Then

$$w(h_1, \ldots, h_n) = 1$$

in H.

Observe that by Lemma 8.2.1 the chain $w(\tau_1, \ldots, \tau_n)$ is a cycle, so we can talk about its homology class.

Proof. First assume $k = 0$. Let $p_i \in \{1, \ldots, 6\}$, $d_i \in \{0, 1\}$ and $e_i \in \{0, 1, 2, 3\}$ be such that

$$\tau_i = \mathrm{flip}^{d_i}(\mathrm{turn}^{e_i}(\tau_{p_i}^{\flat}))$$

as maps $\mathbf{I}^2 \to \mathbf{I}^3$. Let $a_i^{\flat} := [\partial(\sigma_i^{\flat}, \tau_i^{\flat})]$ and $h_i^{\flat} := \mathrm{MI}(\alpha, \beta \mid \sigma_i^{\flat}, \tau_i^{\flat})$, in the notation of Definition 8.1.5. There are words $v_i(t) \in \mathrm{Wrd}^{\pm 1}(t_1, \ldots, t_6)$ such that

$$[\sigma_i] * [\sigma_{p_i}^{\flat}]^{-1} = v_i(a_1^{\flat}, \ldots, a_6^{\flat})$$

in the group $\pi_1(\mathrm{sk}_1 \, \mathrm{sd}^0 \, \mathbf{I}^3)$. According to Corollaries 7.5.2 and 6.3.2, repeated, we have

$$h_i = \mathrm{Ad}_H\big(v_i(h_1^{\flat}, \ldots, h_6^{\flat})\big)\big((h_{p_i}^{\flat})^{(-1)^{d_i}}\big).$$

Define

$$u_i(t) := \mathrm{Ad}_{\mathrm{Wrd}^{\pm 1}(t)}(v_i(t))(y_{p_i}^{(-1)^{d_i}}) \in \mathrm{Wrd}^{\pm 1}(t)$$

using the conjugation operation from (7.1.1). Then $h_i = u_i(h_1^{\flat}, \ldots, h_6^{\flat})$ and $a_i = u_i(a_1^{\flat}, \ldots, a_6^{\flat})$.

Let

$$u(t) := w\big(u_1(t), \ldots, u_n(t)\big) \in \mathrm{Wrd}^{\pm 1}(t).$$

We know that $u(a_1^{\flat}, \ldots, a_6^{\flat}) = 1$ in $\pi_1(\mathrm{sk}_1 \, \mathrm{sd}^0 \, \mathbf{I}^3)$. Hence by Lemmas 8.2.2(1) and 7.1.2 we have

$$u(t) \sim_{\mathrm{can}} \mathrm{Ad}_{\mathrm{Wrd}^{\pm 1}(t)}(v(t))\big((y_1 \cdots y_6)^e\big)$$

in $\mathrm{Wrd}^{\pm 1}(t)$, for some word $v(t)$ and integer e. Passing to the abelian group $H_2(\mathrm{sk}_2 \, \mathrm{sd}^0 \, \mathbf{I}^3)$, with additive notation, we get

$$e \cdot [\tau_1^{\flat} + \cdots + \tau_6^{\flat}] = [u(\tau_1^{\flat}, \ldots, \tau_6^{\flat})] = [w(\tau_1, \ldots, \tau_n)] = 0.$$

Using Lemma 8.2.2(2) we conclude that $e = 0$, and hence $u(t) \sim_{\text{can}} 1$ in $\text{Wrd}^{\pm 1}(t)$. Finally, evaluating in the group H we get

$$w(h_1, \ldots, h_n) = u(h_1^b, \ldots, h_6^b) = 1.$$

For the case $k = 1$ the proof is the same, only using Lemma 8.2.3 instead of Lemma 8.2.2, and working with the with words in the monoid $\text{Wrd}^{\pm 1}(t_1, \ldots, t_{48})$ instead of $\text{Wrd}^{\pm 1}(t_1, \ldots, t_6)$. ◻

8.5 Estimates

We continue with the setup of Section 8.4.

Recall the constants $c_{2'}(\alpha, \beta)$ and $\epsilon_{2'}(\alpha, \beta)$ from Proposition 4.5.4. Among other things, these numbers satisfy $c_{2'}(\alpha, \beta) \geq 1$ and $0 < \epsilon_{2'}(\alpha, \beta) < 1$.

Lemma 8.5.1. *There are constants $c_6(\alpha, \beta)$ and $\epsilon_6(\alpha, \beta)$ with these properties:*

(i) $c_6(\alpha, \beta) \geq c_{2'}(\alpha, \beta)$ *and*

$$0 < \epsilon_6(\alpha, \beta) \leq \min\left(\tfrac{1}{6} \cdot c_6(\alpha, \beta)^{-1} \cdot \epsilon_0(H),\ \epsilon_{2'}(\alpha, \beta)\right).$$

(ii) *Suppose (σ, τ) is a cubical balloon $(\mathbf{I}^3, v_0)$ such that $\text{side}(\tau) < \epsilon_6(\alpha, \beta)$ and $\text{len}(\sigma) \leq 6$. Then*

$$\text{MI}(\alpha, \beta \,|\, \partial(\sigma, \tau)) \in V_0(H)$$

and

$$\left\| \log_H\left(\text{MI}(\alpha, \beta \,|\, \partial(\sigma, \tau))\right) \right\| \leq c_6(\alpha, \beta) \cdot \text{side}(\tau)^3.$$

Proof. The proof is basically the same as that of Lemma 6.1.2, but using Proposition 4.5.4 instead of Proposition 3.5.11. ◻

Suppose (σ, τ) is a nondegenerate cubical balloon in $(\mathbf{I}^3, v_0)$. Let us write $\epsilon := \text{side}(\tau)$; $Z := \tau(\mathbf{I}^3)$, which is an oriented cube in $\mathbf{I}^3$; and $z_0 := \tau(\tfrac{1}{2}, \tfrac{1}{2}, \tfrac{1}{2})$, the midpoint of Z. Let $s = (s_1, s_2, s_3)$ be the orthonormal linear coordinate system on Z such that $\tau^*(s_i) = \epsilon \cdot t_i$.

Assume that the forms $\alpha|_Z$ and $\beta|_Z$ are smooth. Let $\tilde{\alpha}_1, \tilde{\alpha}_2, \tilde{\alpha}_3 \in \mathcal{O}(Z) \otimes \mathfrak{g}$ be the coefficients of $\alpha|_Z$ relative to s, namely

$$\alpha|_Z = \sum_{1 \leq i \leq 3} \tilde{\alpha}_i \cdot \text{d}s_i.$$

And let $\tilde{\beta}_{1,2}, \tilde{\beta}_{1,3}, \tilde{\beta}_{2,3} \in \mathcal{O}(Z) \otimes \mathfrak{h}$ be the coefficients of $\beta|_Z$ relative to s, namely

$$\beta|_Z = \sum_{1 \leq i < j \leq 3} \tilde{\beta}_{i,j} \cdot \text{d}s_i \wedge \text{d}s_j.$$

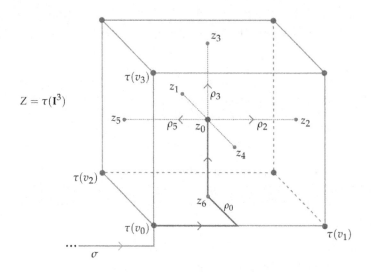

FIGURE 32. Illustration for Section 8.5.

Recall the Lie algebra map

$$\psi_{\mathfrak{h}} = \mathrm{Lie}(\Psi_{\mathfrak{h}}) : \mathfrak{g} \to \mathrm{End}(\mathfrak{h}).$$

For any i we get a function

$$\psi_{\mathfrak{h}}(\tilde{\alpha}_i) \in \mathcal{O}(Z) \otimes \mathrm{End}(\mathfrak{h});$$

so for any i, j, k there is a function

$$\psi_{\mathfrak{h}}(\tilde{\alpha}_i)(\tilde{\beta}_{j,k}) \in \mathcal{O}(Z) \otimes \mathfrak{h}.$$

Define the function
(8.5.2)
$$\tilde{\gamma} := \frac{\partial}{\partial s_1}(\tilde{\beta}_{2,3}) - \frac{\partial}{\partial s_2}(\tilde{\beta}_{1,3}) + \frac{\partial}{\partial s_3}(\tilde{\beta}_{1,2})$$
$$+ \psi_{\mathfrak{h}}(\tilde{\alpha}_1)(\tilde{\beta}_{2,3}) - \psi_{\mathfrak{h}}(\tilde{\alpha}_2)(\tilde{\beta}_{1,3}) + \psi_{\mathfrak{h}}(\tilde{\alpha}_3)(\tilde{\beta}_{1,2}) \in \mathcal{O}(Z) \otimes \mathfrak{h}.$$

Let σ_{pr} be the following string in Z:

(8.5.3) $\quad \sigma_{\mathrm{pr}} := \tau \circ \left(v_0, (\frac{1}{2}, 0, 0)\right) * \left((\frac{1}{2}, 0, 0), (\frac{1}{2}, \frac{1}{2}, 0)\right) * \left((\frac{1}{2}, \frac{1}{2}, 0), (\frac{1}{2}, \frac{1}{2}, \frac{1}{2})\right).$

So σ_{pr} has initial point $\tau(v_0)$ and terminal point z_0; see Figure 32. Define

(8.5.4) $\qquad\qquad g_0 := \mathrm{MI}(\alpha \mid \sigma_{\mathrm{pr}}) \quad \text{and} \quad g := \mathrm{MI}(\alpha \mid \sigma)$

in G.

For $i \in \{1, \ldots, 6\}$ we consider the points $z_i \in Z$ defined as follows:

$$z_1 := \tau(\tfrac{1}{2}, 1, \tfrac{1}{2}), \ z_2 := \tau(1, \tfrac{1}{2}, \tfrac{1}{2}), \ z_3 := \tau(\tfrac{1}{2}, \tfrac{1}{2}, 1),$$

$$z_4 := \tau(\tfrac{1}{2}, 0, \tfrac{1}{2}), \ z_5 := \tau(0, \tfrac{1}{2}, \tfrac{1}{2}), \ z_6 := \tau(\tfrac{1}{2}, \tfrac{1}{2}, 0).$$

See Figure 32. We define the strings

$$(8.5.5) \qquad \rho'_i := (\tau \circ \sigma_i^{\flat}) * (\tau \circ \tau_i^{\flat} \circ \sigma_{\mathrm{pr}}),$$

where σ_{pr} is the probe in $\mathbf{I}^2$; see formula (4.2.4). So ρ'_i has initial point $\tau(v_0)$ and terminal point z_i. Let

$$g'_i := \mathrm{MI}(\alpha \,|\, \rho'_i) \in G.$$

For $i \in \{1, \ldots, 6\}$ we define elements $\lambda_i \in \mathfrak{h}$ as follows:

$$(8.5.6) \qquad \begin{aligned}
\lambda_1 &:= -\epsilon^2 \cdot \Psi_{\mathfrak{h}}(g \cdot g'_1)(\tilde{\beta}_{1,3}(z_1)), \\
\lambda_2 &:= \epsilon^2 \cdot \Psi_{\mathfrak{h}}(g \cdot g'_2)(\tilde{\beta}_{2,3}(z_2)), \\
\lambda_3 &:= \epsilon^2 \cdot \Psi_{\mathfrak{h}}(g \cdot g'_3)(\tilde{\beta}_{1,2}(z_3)), \\
\lambda_4 &:= -\epsilon^2 \cdot \Psi_{\mathfrak{h}}(g \cdot g'_4)(\tilde{\beta}_{1,3}(z_4)), \\
\lambda_5 &:= \epsilon^2 \cdot \Psi_{\mathfrak{h}}(g \cdot g'_5)(\tilde{\beta}_{2,3}(z_5)), \\
\lambda_6 &:= \epsilon^2 \cdot \Psi_{\mathfrak{h}}(g \cdot g'_6)(\tilde{\beta}_{1,2}(z_6)).
\end{aligned}$$

It is easy to see from Definition 4.3.3 that

$$(8.5.7) \qquad \exp_H(\lambda_i) = \mathrm{RP}_0(\alpha, \beta \,|\, \partial_i(\sigma, \tau)).$$

Lemma 8.5.8. *There are constants* $c_7(\alpha, \beta)$ *and* $\epsilon_7(\alpha, \beta)$ *with these properties:*

(i) $c_7(\alpha, \beta) \geq c_6(\alpha, \beta)$ *and*

$$0 < \epsilon_7(\alpha, \beta) \leq \min\left(\epsilon_6(\alpha, \beta), \tfrac{1}{10}\epsilon_1(\alpha)\right).$$

(ii) *Let* (σ, τ) *be a nondegenerate cubical balloon in* $(\mathbf{I}^3, v_0)$ *with* $\epsilon :=$ *side*(τ). *Assume that* $\epsilon < \epsilon_7(\alpha, \beta)$, *len*$(\sigma) \leq 6$, *and the forms* $\alpha|_Z$ *and* $\beta|_Z$ *are smooth. Then, with* λ_i *as in (8.5.6), the following estimate holds:*

$$\left\| \log_H(\mathrm{MI}(\alpha, \beta \,|\, \partial(\sigma, \tau))) - \sum_{i=1}^{6} \lambda_i \right\| \leq c_7(\alpha, \beta) \cdot \epsilon^4.$$

Proof. Let us set

$$c := 12 \cdot \max\left(c_2(\alpha, \beta), \ \exp(c_4(\alpha, \Psi_{\mathfrak{h}}) \cdot 9) \cdot \|\beta\|_{\mathrm{Sob}}\right).$$

According to Proposition 4.5.2(2), if $\epsilon < \epsilon_2(\alpha, \beta)$ then

$$(8.5.9) \qquad \left\| \log_H(\mathrm{MI}(\alpha, \beta \,|\, \partial_i(\sigma, \tau))) - \lambda_i \right\| \leq c_2(\alpha, \beta) \cdot \epsilon^4 \leq \tfrac{1}{12}c \cdot \epsilon^4.$$

And by Proposition 3.5.10 we know that

$$\|\lambda_i\| \le \exp\big(c_4(\alpha, \Psi_\mathfrak{h}) \cdot 9\big) \cdot \|\beta\|_{\mathrm{Sob}} \cdot \epsilon^2 \le \tfrac{1}{12} c \cdot \epsilon^2 .$$

Hence

(8.5.10) $$\|\log_H\big(\mathrm{MI}(\alpha, \beta \,|\, \partial_i(\sigma, \tau))\big)\| \le \tfrac{1}{6} c \cdot \epsilon^2 .$$

We will take

$$\epsilon_7(\alpha, \beta) := \min\big(\epsilon_6(\alpha, \beta), c^{-1/2} \cdot \epsilon_0(H)^{1/2}\big) .$$

Now assume that $\epsilon < \epsilon_7(\alpha, \beta)$. Then by property (ii) of Theorem 2.1.2, used in conjunction with the bound (8.5.10), we get

$$\|\log_H\big(\textstyle\prod_{i=1}^6 \mathrm{MI}(\alpha, \beta \,|\, \partial_i(\sigma, \tau))\big) - \sum_{i=1}^6 \mathrm{MI}(\alpha, \beta \,|\, \partial_i(\sigma, \tau))\|$$
$$\le c_0(H) \cdot (c \cdot \epsilon^2)^2 = c_0(H) \cdot c^2 \cdot \epsilon^4 .$$

Combining this with (8.5.9) we obtain

(8.5.11) $$\|\log_H\big(\textstyle\prod_{i=1}^6 \mathrm{MI}(\alpha, \beta \,|\, \partial_i(\sigma, \tau))\big) - \sum_{i=1}^6 \lambda_i\|$$
$$\le \big(c_0(H) \cdot c^2 + \tfrac{1}{2} c\big) \cdot \epsilon^4 .$$

Thus the constant

$$c_7(\alpha, \beta) := \max\big(c_0(H) \cdot c^2 + \tfrac{1}{2} c, c_6(\alpha, \beta)\big)$$

works. □

Definition 8.5.12. (1) Let (σ, τ) be a square kite in $(\mathbf{I}^3, v_0)$. We will say that (σ, τ) is (α, β)-*tiny* (in this chapter) if $\mathrm{side}(\tau) < \epsilon_7(\alpha, \beta)$ and $\mathrm{len}(\sigma) \le 9$.

(2) Let (σ, τ) be a cubical balloon in $(\mathbf{I}^3, v_0)$. We will say that (σ, τ) is (α, β)-*tiny* if $\mathrm{side}(\tau) < \epsilon_7(\alpha, \beta)$ and $\mathrm{len}(\sigma) \le 6$.

Note that if (σ, τ) is an (α, β)-tiny cubical balloon, then the kites $\partial_i(\sigma, \tau)$ are all (α, β)-tiny.

Let (σ, τ) be a cubical balloon in $(\mathbf{I}^2, v_0)$. For $i \in \{1, \ldots, 6\}$ let ρ_i be the linear map $\mathbf{I}^1 \to Z$ such that $\rho_i(v_0) = z$ and $\rho_i(v_1) = z_i$. Recall the string σ_{pr} from (8.5.3). For every $i \in \{1, \ldots, 6\}$ the string $\sigma_{\mathrm{pr}} * \rho_i$ has initial point $\tau(v_0)$ and terminal point z_i. Let

$$g_i := \mathrm{MI}(\alpha \,|\, \rho_i) \in G.$$

The form $\alpha|_Z$ is smooth, and hence we have

$$\alpha' := \psi_\mathfrak{h}(\alpha|_Z) \in \Omega^1(Z) \otimes \mathrm{End}(\mathfrak{h}).$$

As in Definition 1.3.5 there is an associated constant form at z_0:

$$\alpha'(z_0) \in \Omega^1_{\mathrm{const}}(Z) \otimes \mathrm{End}(\mathfrak{h}).$$

In this way for every i we get

$$\int_{\rho_i} \alpha'(z_0) \in \mathrm{End}(\mathfrak{h}).$$

Lemma 8.5.13. *There is a constant $c_8(\alpha, \beta)$ with this property:*

(∗) *Let (σ, τ) be a nondegenerate (α, β)-tiny cubical balloon in $(\mathbf{I}^3, v_0)$, with $\epsilon := \mathrm{side}(\tau)$. Then, in the notation above, the estimate*

$$\left\| \Psi_{\mathfrak{h}}(g \cdot g_i') - \Psi_{\mathfrak{h}}(g \cdot g_0) \circ \left(\mathbf{1} + \int_{\rho_i} \alpha'(z_0)\right) \right\| \leq c_8(\alpha, \beta) \cdot \epsilon^2$$

holds for every $i \in \{1, \ldots, 6\}$.

Proof. Consider such a balloon. Recall the strings ρ_i' from formula (8.5.5). Since $\epsilon < \epsilon_7(\alpha, \beta) \leq \frac{1}{10}\epsilon_1(\alpha)$, we see that for every i the closed string $\rho_i' * (\sigma_{\mathrm{pr}} * \rho_i)^{-1}$ has

$$\mathrm{len}(\rho_i' * (\sigma_{\mathrm{pr}} * \rho_i)^{-1}) < 10 \cdot \epsilon \leq \epsilon_1(\alpha).$$

And this closed string bounds an area less than $10 \cdot \epsilon^2$. Hence by Corollary 3.5.12 we have

$$\left\| \log_G\left(g_i' \cdot (g_0 \cdot g_i)^{-1}\right) \right\| \leq c_0(G) \cdot c_1(\alpha)^2 \cdot (10 \cdot \epsilon)^2 + (10 \cdot \epsilon^2) \cdot \|\alpha\|_{\mathrm{Sob}}.$$

Therefore there is a constant c', depending only on α, such that

(8.5.14) $$\left\| \Psi_{\mathfrak{h}}(g_i') - \Psi_{\mathfrak{h}}(g_0 \cdot g_i) \right\| \leq c' \cdot \epsilon^2.$$

By Proposition 3.5.10 we know that

$$\|\Psi_{\mathfrak{h}}(g)\| \leq \exp\left(c_4(\alpha, \Psi_{\mathfrak{h}}) \cdot 9\right) \cdot \|\alpha\|_{\mathrm{Sob}}$$

and

$$\|\Psi_{\mathfrak{h}}(g_0)\| \leq \exp\left(c_4(\alpha, \Psi_{\mathfrak{h}}) \cdot 2\epsilon\right) \cdot \|\alpha\|_{\mathrm{Sob}}.$$

Therefore

$$\left\| \Psi_{\mathfrak{h}}(g \cdot g_i') - \Psi_{\mathfrak{h}}(g \cdot g_0 \cdot g_i) \right\| = \left\| \Psi_{\mathfrak{h}}(g) \circ \left(\Psi_{\mathfrak{h}}(g_i') - \Psi_{\mathfrak{h}}(g_0 \cdot g_i)\right) \right\|$$

$$\leq \left\| \Psi_{\mathfrak{h}}(g) \right\| \cdot \left\| \Psi_{\mathfrak{h}}(g_i') - \Psi_{\mathfrak{h}}(g_0 \cdot g_i) \right\|$$

$$\leq \exp\left(c_4(\alpha, \Psi_{\mathfrak{h}}) \cdot 9\right) \cdot \|\alpha\|_{\mathrm{Sob}} \cdot c' \cdot \epsilon^2.$$

On the other hand, by Proposition 3.5.13 we have the estimate

$$\left\| \Psi_{\mathfrak{h}}(g_i) - \left(\mathbf{1} + \int_{\rho_i} \alpha'(z_0)\right) \right\| \leq c_5(\alpha, \Psi_{\mathfrak{h}}) \cdot \tfrac{1}{4}\epsilon^2.$$

We conclude that the constant

$$c_8(\alpha, \beta) := \exp\left(c_4(\alpha, \Psi_{\mathfrak{h}}) \cdot 9\right) \cdot \|\alpha\|_{\mathrm{Sob}} \cdot c'$$

$$+ c_5(\alpha, \Psi_{\mathfrak{h}}) \cdot \tfrac{1}{4} \cdot \exp\left(c_4(\alpha, \Psi_{\mathfrak{h}}) \cdot 12\right) \cdot \|\alpha\|_{\mathrm{Sob}}^2$$

works.   □

Lemma 8.5.15. *There is a constant $c_9(\alpha, \beta)$ with this property:*

(∗) *Let (σ, τ) be a nondegenerate (α, β)-tiny cubical balloon in $(\mathbf{I}^3, v_0)$, with $\epsilon := \mathrm{side}(\tau)$. Then, in the notation above, the following estimate holds:*

$$\left\| \textstyle\sum_{i=1}^{6} \lambda_i - \epsilon^3 \cdot \Psi_\mathfrak{h}(g \cdot g_0)(\tilde{\gamma}(z_0)) \right\| \leq \epsilon^4 \cdot c_9(\alpha, \beta).$$

Proof. The strategy is to try to estimate the elements λ_i.

We begin with $i = 1$. Define

$$\mu_1 := \epsilon^2 \cdot \Psi_\mathfrak{h}(g \cdot g_0)\Big(-\tilde{\beta}_{1,3}(z_0) - \tfrac{1}{2} \cdot \epsilon \cdot \psi_\mathfrak{h}(\tilde{\alpha}_2(z_0))(\tilde{\beta}_{1,3}(z_0))$$

$$- \tfrac{1}{2} \cdot \epsilon \cdot (\tfrac{\partial}{\partial s_2}\tilde{\beta}_{1,3})(z_0) \Big).$$

The Taylor expansion to first order of the function $\tilde{\beta}_{1,3}$ around z_0 gives us

(8.5.16) $\left\| \tilde{\beta}_{1,3}(z_1) - (\tilde{\beta}_{1,3}(z_0) - \tfrac{1}{2}\epsilon \cdot (\tfrac{\partial}{\partial s_2}\tilde{\beta}_{1,3})(z_0)) \right\| \leq \tfrac{1}{4}\epsilon^2 \cdot \|\beta\|_{\mathrm{Sob}}$.

And Lemma 8.5.13, for $i = 1$, gives

(8.5.17) $\left\| \Psi_\mathfrak{h}(g \cdot g_i') - \Psi_\mathfrak{h}(g \cdot g_0) \circ (1 - \tfrac{1}{2}\epsilon \cdot \tilde{\alpha}_2(z_0)) \right\| \leq c_8(\alpha, \beta) \cdot \epsilon^2$.

Let

$$d := \exp(c_4(\alpha, \Psi_\mathfrak{h}) \cdot 9) \cdot \tfrac{1}{4} \cdot \|\beta\|_{\mathrm{Sob}} + c_8(\alpha, \beta) \cdot 2 \cdot \|\beta\|_{\mathrm{Sob}}$$

$$+ \exp(c_4(\alpha, \Psi_\mathfrak{h}) \cdot 9) \cdot \tfrac{1}{2} \cdot \|\alpha'\|_{\mathrm{Sob}} \cdot \|\beta\|_{\mathrm{Sob}}$$.

By combining the estimates (8.5.16) and (8.5.17) we obtain

(8.5.18) $\|\lambda_1 - \mu_1\| \leq d \cdot \epsilon^4$.

We do the same thing for λ_4: let

$$\mu_4 := \epsilon^2 \cdot \Psi_\mathfrak{h}(g \cdot g_0)\Big(\tilde{\beta}_{1,3}(z_0) - \tfrac{1}{2} \cdot \epsilon \cdot \psi_\mathfrak{h}(\tilde{\alpha}_2(z_0))(\tilde{\beta}_{1,3}(z_0))$$

$$- \tfrac{1}{2} \cdot \epsilon \cdot (\tfrac{\partial}{\partial s_2}\tilde{\beta}_{1,3})(z_0) \Big).$$

The same sort of calculation that got us (8.5.18), now gives the inequality

(8.5.19) $\|\lambda_4 - \mu_4\| \leq d \cdot \epsilon^4$.

And clearly

$$\mu_1 + \mu_4 = \epsilon^3 \cdot \Psi_\mathfrak{h}(g \cdot g_0)\big(-\psi_\mathfrak{h}(\tilde{\alpha}_2(z_0))(\tilde{\beta}_{1,3}(z_0)) - (\tfrac{\partial}{\partial s_2}\tilde{\beta}_{1,3})(z_0) \big).$$

Plugging inequalities (8.5.18) and (8.5.19) into this, we get
(8.5.20)

$$\left\| (\lambda_1 + \lambda_4) - \big(\epsilon^3 \cdot \Psi_\mathfrak{h}(g \cdot g_0)\big(-\psi_\mathfrak{h}(\tilde{\alpha}_2(z_0))(\tilde{\beta}_{1,3}(z_0)) - (\tfrac{\partial}{\partial s_2}\tilde{\beta}_{1,3})(z_0) \big) \big) \right\|$$

$$\leq \epsilon^4 \cdot 2d .$$

By a similar calculation for the pairs of indices $(2,5)$ and $(3,6)$ we get these analogues of formula $(8.5.20)$:

$$\| (\lambda_2 + \lambda_5) - (\epsilon^3 \cdot \Psi_\mathfrak{h}(g \cdot g_0)(\psi_\mathfrak{h}(\tilde{\alpha}_1(z_0))(\tilde{\beta}_{2,3}(z_0)) + (\tfrac{\partial}{\partial s_1}\tilde{\beta}_{2,3})(z_0))) \|$$
$$\leq \epsilon^4 \cdot 2d$$

and

$$\| (\lambda_3 + \lambda_6) - (\epsilon^3 \cdot \Psi_\mathfrak{h}(g \cdot g_0)(\psi_\mathfrak{h}(\tilde{\alpha}_3(z_0))(\tilde{\beta}_{1,2}(z_0)) + (\tfrac{\partial}{\partial s_3}\tilde{\beta}_{1,2})(z_0))) \|$$
$$\leq \epsilon^4 \cdot 2d .$$

Thus, plugging in the value of $\tilde{\gamma}(z_0)$ from $(8.5.2)$ we get

$$\| \textstyle\sum_{i=1}^{6} \lambda_i - \epsilon^3 \cdot \Psi_\mathfrak{h}(g \cdot g_0)(\tilde{\gamma}(z_0)) \| \leq \epsilon^4 \cdot 6d .$$

To finish we take $c_8(\alpha,\beta) := 6d$. $\qquad\square$

The last two lemmas combined give us:

Lemma 8.5.21. *Let (σ,τ) be a nondegenerate (α,β)-tiny cubical balloon in $(\mathbf{I}^3, v_0)$, with $\epsilon := \mathrm{side}(\tau)$, $Z := \tau(\mathbf{I}^3)$ and $z := (\tfrac{1}{2},\tfrac{1}{2},\tfrac{1}{2})$. Assume that $\alpha|_Z$ and $\beta|_Z$ are smooth. Let $\tilde{\gamma} \in \mathcal{O}(Z) \otimes \mathfrak{h}$ be the function from equation $(8.5.2)$, and let $g, g_0 \in G$ be the group elements from equation $(8.5.4)$. Then*

$$\| \log_H(\mathrm{MI}(\alpha,\beta\,|\,\partial(\sigma,\tau))) - \epsilon^3 \cdot \Psi_\mathfrak{h}(g \cdot g_0)(\tilde{\gamma}(z_0)) \| \leq \epsilon^4 \cdot c_{10}(\alpha,\beta) ,$$

where

$$c_{10}(\alpha,\beta) := c_8(\alpha,\beta) + c_9(\alpha,\beta).$$

If, moreover, $\tilde{\gamma}$ happens to be an inert function, then

$$\Psi_\mathfrak{h}(g \cdot g_0)(\tilde{\gamma}(z_0)) = \Psi_{\mathfrak{h},\alpha}(\tilde{\gamma})(z_0),$$

and therefore

$$\| \log_H(\mathrm{MI}(\alpha,\beta\,|\,\partial(\sigma,\tau))) - \epsilon^3 \cdot \Psi_{\mathfrak{h},\alpha}(\tilde{\gamma})(z_0) \| \leq \epsilon^4 \cdot c_{10}(\alpha,\beta) .$$

8.6 Stokes Theorem

Here again we are in the general situation: (X, x_0) is a pointed polyhedron,

$$\mathbf{C}/X = (G, H, \Psi, \Phi_0, \Phi_X)$$

is a Lie quasi crossed module with additive feedback, and (α,β) is a piecewise smooth connection-curvature pair in $\mathbf{C}/X$.

The Lie group map $\Psi_\mathfrak{h} : G \to \mathrm{GL}(\mathfrak{h})$ induces a Lie algebra map

$$\psi_\mathfrak{h} := \mathrm{Lie}(\Psi_\mathfrak{h}) : \mathfrak{g} \to \mathfrak{gl}(\mathfrak{h}) = \mathrm{End}(\mathfrak{h}).$$

By tensoring with $\Omega_{\mathrm{pws}}(X)$ this induces a map of DG Lie algebras

$$\psi_{\mathfrak{h}} : \Omega_{\mathrm{pws}}(X) \otimes \mathfrak{g} \to \Omega_{\mathrm{pws}}(X) \otimes \mathrm{End}(\mathfrak{h}).$$

In this way from the pair $\alpha \in \Omega^1_{\mathrm{pws}}(X) \otimes \mathfrak{g}$ and $\beta \in \Omega^2_{\mathrm{pws}}(X) \otimes \mathfrak{h}$ we get

$$\psi_{\mathfrak{h}}(\alpha)(\beta) \in \Omega^3_{\mathrm{pws}}(X) \otimes \mathfrak{h}.$$

Definition 8.6.1. Let (X, x_0) be a pointed polyhedron, let

$$\mathbf{C}/X = (G, H, \Psi, \Phi_0, \Phi_X)$$

be a Lie quasi crossed module with additive feedback, and let (α, β) be a piecewise smooth connection-curvature pair in $\mathbf{C}/X$. The 3-*curvature of* (α, β) is the form

$$\gamma := \mathrm{d}(\beta) + \psi_{\mathfrak{h}}(\alpha)(\beta) \in \Omega^3_{\mathrm{pws}}(X) \otimes \mathfrak{h}.$$

Recall the notion of orientation of a polyhedron (Section 1.8). If Z is an oriented cube, and (s_1, s_2, s_3) is a positively oriented orthonormal linear coordinate system on Z, then

$$\mathrm{or}(Z) = \mathrm{d}s_1 \wedge \mathrm{d}s_2 \wedge \mathrm{d}s_3.$$

Lemma 8.6.2. *Assume that* $(X, x_0) = (\mathbf{I}^3, v_0)$, *so that we are in the situation of Section 8.5. Let* $Z \subset \mathbf{I}^3$ *be an oriented nondegenerate cube, such that* $\alpha|_Z$ *and* $\beta|_Z$ *are smooth. Let* $\tilde{\gamma} \in \mathcal{O}(Z) \otimes \mathfrak{h}$ *be the function from formula (8.5.2). Then* $\tilde{\gamma}$ *is the coefficient of* $\gamma|_Z$; *namely*

$$\gamma|_Z = \tilde{\gamma} \cdot \mathrm{or}(Z).$$

Proof. This amounts to expanding the formula in Definition 8.6.1 into coordinates. $\qquad\square$

Lemma 8.6.3. *Let* (σ, τ) *be a balloon in* (X, x_0). *Then*

$$\mathrm{MI}(\alpha, \beta \,|\, \partial(\sigma, \tau)) \in H_0.$$

Proof. By definition we have

$$\mathrm{MI}(\alpha, \beta \,|\, \partial(\sigma, \tau)) = \prod_{i=1,\dots,6} \mathrm{MI}(\alpha, \beta \,|\, (\sigma, \tau) \circ (\sigma_i^{\flat}, \tau_i^{\flat}))$$

in the notation of Definition 8.1.5. For every index i we have, by Theorem 6.2.1, an equality

$$\Phi_0 \big(\mathrm{MI}(\alpha, \beta \,|\, (\sigma, \tau) \circ (\sigma_i^{\flat}, \tau_i^{\flat})) \big) = \mathrm{MI}(\alpha \,|\, \partial((\sigma, \tau) \circ (\sigma_i^{\flat}, \tau_i^{\flat})))$$

in the group G. Now from Figure 31 we see that the closed string

$$\partial\partial\mathbf{I}^3 = \prod_{i=1,\dots,6} \partial(\sigma_i^{\flat}, \tau_i^{\flat})$$

is cancellation equivalent to the empty string in the monoid of strings in $\mathbf{I}^3$. Therefore the closed string

$$\partial\partial(\sigma,\tau) = \prod_{i=1,\dots,6} \partial((\sigma,\tau) \circ (\sigma_i^{\flat}, \tau_i^{\flat}))$$

is cancellation equivalent to the empty string in the monoid of strings in X. By Proposition 3.5.8 we can conclude that

$$\prod_{i=1,\dots,6} \mathrm{MI}(\alpha, \beta \mid \partial((\sigma,\tau) \circ (\sigma_i^{\flat}, \tau_i^{\flat}))) = 1$$

in G. Hence

$$\Phi_0\big(\mathrm{MI}(\alpha, \beta \mid \partial(\sigma,\tau))\big) = 1.$$

$\square$

Lemma 8.6.4. *Let* (σ,τ) *be a balloon in* (X, x_0). *Then there is a piecewise linear map* $f : (\mathbf{I}^3, v_0) \to (X, x_0)$, *and a cubical balloon* (σ', τ') *in* $(\mathbf{I}^3, v_0)$, *such that* $\mathrm{len}(\sigma') \leq 2$, $f|_{\tau'(\mathbf{I}^3)}$ *is linear, and*

$$(\sigma,\tau) = f \circ (\sigma', \tau')$$

as balloons in (X, x_0).

Proof. Just like the proof of Proposition 4.1.6. $\square$

Lemma 8.6.5. *Let* (σ,τ) *be a balloon in* (X, x_0). *Take any* $k \geq 0$. *Then*

$$\mathrm{MI}(\alpha, \beta \mid \partial(\mathrm{tes}^k(\sigma,\tau))) = \mathrm{MI}(\alpha, \beta \mid \partial(\sigma,\tau)).$$

Proof. Due to the recursive nature of the tessellations it is enough to consider the case $k = 1$. As in the first paragraph in the proof of Lemma 7.3.3, and using Lemma 8.6.4, we can assume that $(X, x_0) = (\mathbf{I}^2, v_0)$ and $(\sigma,\tau) = (\sigma_1^0, \tau_1^0)$, the basic balloon.

The assertion is now an easy consequence of Lemma 8.4.2 and Corollary 7.5.2. $\square$

Theorem 8.6.6 (Nonabelian 3-dimensional Stokes Theorem). *Let* (X, x_0) *be a pointed polyhedron, let* $\mathbf{C}/X$ *be a Lie quasi crossed module with additive feedback, let* (α, β) *be a piecewise smooth connection-curvature pair in* $\mathbf{C}/X$, *and let* γ *be the 3-curvature of* (α, β). *Then:*

(1) *The form* γ *is inert.*

(2) *For any balloon* (σ,τ) *in* (X, x_0) *one has*

$$\mathrm{MI}(\alpha, \beta \mid \partial(\sigma,\tau)) = \mathrm{MI}(\alpha, \gamma \mid \tau)$$

in H.

Note that assertion (1) of the theorem is a *generalized Bianchi identity*.

Proof. (1) Since γ is a 3-form, it is enough to prove that the form $\gamma|_Z \in \Omega^3_{\mathrm{pws}}(Z) \otimes \mathfrak{h}$ is inert for every cube Z in X. Given such a cube Z, choose a linear map $f : \mathbf{I}^3 \to X$ such that $Z = f(\mathbf{I}^3)$. Then $\gamma|_Z$ is inert if and only if $f^*(\gamma) \in \Omega^3_{\mathrm{pws}}(\mathbf{I}^3) \otimes \mathfrak{h}$ is inert, with respect to the induced Lie quasi crossed module with additive feedback $f^*(\mathbf{C}/X)$. Hence we might as well assume that $(X, x_0) = (\mathbf{I}^3, v_0)$.

Now the singular locus of α and the singular locus of β are contained in a finite union of polygons in $\mathbf{I}^3$. So by continuity it is enough to show that $\gamma|_Z$ is inert for cubes $Z \subset \mathbf{I}^3$ such that $\alpha|_Z$ and $\beta|_Z$ are smooth. In this case γ is also smooth.

Given such a cube Z, choose an orientation on it, and let $\tilde{\gamma} \in \mathcal{O}(Z) \otimes \mathfrak{h}$ be the coefficient of $\gamma|_Z$ (in the sense of Definition 1.8.2). We have to prove that the function $\tilde{\gamma}$ is inert; namely that $\tilde{\gamma}(z) \in \mathrm{Ker}(\Phi_X(z))$ for every $z \in Z$. Again by continuity, it is enough to look at $z \in \mathrm{Int}\, Z$.

We are allowed to move the cube Z around the point z and to shrink it. Hence it is enough to take an (α, β)-tiny balloon (σ, τ), and to show that $\tilde{\gamma}(z) \in \mathrm{Ker}(\Phi_X(z))$ for the midpoint $z := \tau(\frac{1}{2}, \frac{1}{2}, \frac{1}{2})$ of $Z := \tau(\mathbf{I}^3)$.

Let g and g_0 be the group elements from equation (8.5.4). By Lemma 8.6.3 we have

$$\mathrm{MI}(\alpha, \beta \,|\, \partial(\sigma, \tau) \in H_0,$$

and by Lemma 8.5.1 we have

$$\mathrm{MI}(\alpha, \beta \,|\, \partial(\sigma, \tau)) \in V_0(H).$$

Taking logarithms we see that

$$\log_H(\mathrm{MI}(\alpha, \beta \,|\, \partial(\sigma, \tau)) \in \mathfrak{h}_0.$$

By Lemma 8.5.21, it follows that the distance of the element $\Psi_{\mathfrak{h}}(g \cdot g_0)(\tilde{\gamma}(z))$ from the linear subspace $\mathfrak{h}_0$ is at most $c_{10} \cdot \epsilon^4$.

Now by Proposition 3.5.10(1) there is a uniform bound on the norm of the operator $\Psi_{\mathfrak{h}}(g \cdot g_0)^{-1}$; say c'. So the distance of $\tilde{\gamma}(z)$ from the subspace $\mathrm{Ker}(\Phi_X(z)) \subset \mathfrak{h}$ is at most $c' \cdot c \cdot \epsilon$.

Since we can make ϵ arbitrarily small, we conclude that $\tilde{\gamma}(z) \in \mathrm{Ker}(\Phi_X(z))$.

(2) By the functoriality of $\mathrm{MI}(\alpha, \beta \,|\, \partial(\sigma, \tau))$ and $\mathrm{MI}(\alpha, \gamma \,|\, \tau)$ (see Propositions 4.5.3 and 8.3.13), and using the construction of Lemma 8.6.4, we can assume that $(X, x_0) = (\mathbf{I}^3, v_0)$ and $\mathrm{len}(\sigma) \leq 2$.

Take any $k \geq 0$. Consider the sequence

$$\mathrm{tes}^k(\sigma, \tau) = ((\sigma, \tau) \circ (\sigma_i^k, \tau_i^k)))_{i=1,\dots,8^k}$$

which was defined in Section 8.1. The twisted multiplicative integration of inert forms is multiplicative:

$$\mathrm{MI}(\alpha, \gamma \,|\, \tau) = \prod_{i=1,\ldots,8^k} \mathrm{MI}(\alpha, \gamma \,|\, \tau \circ \tau_i^k).$$

On the other hand, by Lemma 8.6.5 we know that

$$\mathrm{MI}(\alpha, \beta \,|\, \partial(\sigma, \tau)) = \prod_{i=1,\ldots,8^k} \mathrm{MI}(\alpha, \beta \,|\, \partial((\sigma, \tau) \circ (\sigma_i^k, \tau_i^k))).$$

So it suffices to prove that

$$\mathrm{MI}(\alpha, \beta \,|\, \partial((\sigma, \tau) \circ (\sigma_i^k, \tau_i^k))) = \mathrm{MI}(\alpha, \gamma \,|\, \tau \circ \tau_i^k)$$

for every $i \in \{1, \ldots, 8^k\}$.

If k is large enough then all the balloons in $\mathrm{tes}^k(\sigma, \tau)$ are (α, β)-tiny. We conclude that it suffices to prove the equality

$$\mathrm{MI}(\alpha, \beta \,|\, \partial(\sigma, \tau)) = \mathrm{MI}(\alpha, \gamma \,|\, \tau)$$

for an (α, β)-tiny balloon (σ, τ) in $(\mathbf{I}^3, v_0)$.

Suppose we are given an (α, β)-tiny balloon (σ, τ), and a natural number k. Let $\epsilon := \mathrm{side}(\tau)$. For any index $i \in \{1, \ldots, 8^k\}$ let $Z_i := (\tau \circ \tau_i^k)(\mathbf{I}^3)$, which is an oriented cube in $\mathbf{I}^3$ of side $(\frac{1}{2})^k \cdot \epsilon$, and let $z_i := (\tau \circ \tau_i^k)(\frac{1}{2}, \frac{1}{2}, \frac{1}{2})$. Define

$$\mathrm{good}(\tau, k) := \{i \mid \alpha|_{Z_i} \text{ and } \beta|_{Z_i} \text{ are smooth}\},$$

and define $\mathrm{bad}(\tau, k)$ to be the complement of $\mathrm{good}(\tau, k)$ in $\{1, \ldots, 8^k\}$. Since the singular loci of α and β are contained in a finite union of polygons, it follows that

$$|\mathrm{bad}(\tau, k)| \le a_2(\alpha, \beta) \cdot 4^k + a_0(\alpha, \beta)$$

for some constants $a_0(\alpha, \beta), a_2(\alpha, \beta)$, that are independent of k; cf. Lemma 4.4.13.

For $i \in \mathrm{good}(\tau, k)$ let $\tilde{\gamma}_i \in \mathcal{O}(Z_i) \otimes \mathfrak{h}$ be the coefficient of $\gamma|_{Z_i}$, in the sense of Definition 1.8.2, and let

$$\mu_i := (\tfrac{1}{2})^{3k} \cdot \epsilon^3 \cdot \Psi_{\mathfrak{h},\alpha}(\tilde{\gamma}_i)(z_i) \in \mathfrak{h}_0.$$

According to Lemmas 8.5.21 and 8.6.2 we know that

$$\left\| \log_H(\mathrm{MI}(\alpha, \beta \,|\, \partial((\sigma, \tau) \circ (\sigma_i^k, \tau_i^k)))) - \mu_i \right\| \le (\tfrac{1}{2})^{4k} \cdot \epsilon^4 \cdot c_{10}(\alpha, \beta).$$

On the other hand, for $i \in \mathrm{bad}(\tau, k)$ let $\mu_i := 0 \in \mathfrak{h}$. According to Lemma 8.5.1 we have the inequality

$$\left\| \log_H(\mathrm{MI}(\alpha, \beta \,|\, \partial((\sigma, \tau) \circ (\sigma_i^k, \tau_i^k)))) - \mu_i \right\| \le (\tfrac{1}{2})^{3k} \cdot \epsilon^3 \cdot c_6(\alpha, \beta).$$

Let

$$RS_k(\alpha, \gamma \mid \tau) := \sum_{i=1}^{8^k} \mu_i \in \mathfrak{h}_0.$$

Applying property (iv) of Theorem 2.1.2 with the inequalities above we get

$$\left\| \log_H \left(\mathrm{MI}(\alpha, \beta \mid \sigma, \tau) \right) - RS_k(\alpha, \gamma \mid \tau) \right\|$$
$$\leq c_0(H) \cdot \left(8^k \cdot (\tfrac{1}{2})^{4k} \cdot \epsilon^4 \cdot c_{10}(\alpha, \beta) \right.$$
$$\left. + (a_2(\alpha, \beta) \cdot 4^k + a_0(\alpha, \beta)) \cdot (\tfrac{1}{2})^{3k} \cdot \epsilon^3 \cdot c_6(\alpha, \beta) \right).$$

We see that the difference tends to 0 as $k \to \infty$. But

$$\lim_{k \to \infty} RS_k(\alpha, \gamma \mid \tau) = \log_H \left(\mathrm{MI}(\alpha, \gamma \mid \tau) \right).$$

$\square$

Remark 8.6.7. Part (1) of Theorem 8.6.6 is a variant of the Bianchi identity.

9 Multiplicative Integration on Triangular Kites

In this chapter (X, x_0) is a pointed manifold. Recall that by Convention 1.4.7 this means that X is a smooth manifold with sharp corners (cf. Definition 1.4.2); so it could be a polyhedron. We introduce triangular kites in (X, x_0) and the corresponding multiplicative integration.

9.1 Triangular Kites and Balloons

Recall that the polyhedra $\mathbf{I}^1$ and $\mathbf{\Delta}^1$ are identified via the linear isomorphism $\mathbf{\Delta}^1 \xrightarrow{\simeq} \mathbf{I}^1$ that on vertices is $(v_0, v_1) \mapsto (v_0, v_1)$.

A *piecewise smooth path* in X is a piecewise smooth map $\sigma : \mathbf{I}^1 \to X$. When convenient we shall view such a path as a piecewise smooth map $\mathbf{\Delta}^1 \to X$ using the identification above.

Suppose $\sigma_1, \sigma_2 : \mathbf{I}^1 \to X$ are piecewise smooth paths satisfying $\sigma_1(v_1) = \sigma_2(v_0)$. Their *product* is the piecewise smooth path

$$\sigma_1 * \sigma_2 : \mathbf{I}^1 \to X$$

defined as follows:

$$(\sigma_1 * \sigma_2)(a) := \begin{cases} \sigma_1(2a) & \text{if } 0 \le a \le \frac{1}{2} \\ \sigma_2(2a - 1) & \text{if } \frac{1}{2} \le a \le 1. \end{cases}$$

Note that this is the standard product used in homotopy theory, and it is distinct from the concatenation operation on strings. In particular this product is not associative nor unital.

The *inverse* of a piecewise smooth path $\sigma : \mathbf{I}^1 \to X$ is the piecewise smooth path $\sigma^{-1} : \mathbf{I}^1 \to X$ defined by

$$\sigma^{-1}(a) := \sigma(1 - a).$$

Suppose Z is a polyhedron, $f : Z \to X$ is a piecewise smooth map, and $\sigma = (\sigma_1, \ldots, \sigma_m)$ is a string in Z, with $m \ge 1$. This data gives rise to a piecewise smooth path $f \circ \sigma$ in X defined as follows:

(9.1.1) $\qquad f \circ \sigma := \big((f \circ \sigma_1) * (f \circ \sigma_2)\big) * \cdots * (f \circ \sigma_m).$

In this formula we view each $f \circ \sigma_i$ as a piecewise smooth path in X, and the multiplication $*$ is that of paths. For the empty string σ (i.e. $m = 0$) the path $f \circ \sigma$ is not always defined; but if X has a base point x_0, then we usually define $f \circ \sigma$ to be the constant path x_0.

Definition 9.1.2. A *piecewise smooth triangular kite* in (X, x_0) is a pair (σ, τ), consisting of piecewise smooth maps $\sigma : \mathbf{I}^1 \to X$ and $\tau : \mathbf{\Delta}^2 \to X$, satisfying $\sigma(v_0) = x_0$ and $\sigma(v_1) = \tau(v_0)$.

See Figure 1 for an illustration.

The *boundary* of $\mathbf{\Delta}^2$ is the string

(9.1.3) $\partial \mathbf{\Delta}^2 := (v_0, v_1) * (v_1, v_2) * (v_2, v_0)$

(consisting of 3 pieces) in $\mathbf{\Delta}^2$.

Definition 9.1.4. Let (σ, τ) be a piecewise smooth triangular kite in (X, x_0). Its *boundary* is the piecewise smooth path

$$\partial(\sigma, \tau) := \left(\sigma * (\tau \circ \partial \mathbf{\Delta}^2)\right) * \sigma^{-1}.$$

See Figure 2.

Definition 9.1.5. A *piecewise smooth triangular balloon* in (X, x_0) is a pair (σ, τ), consisting of piecewise smooth maps $\sigma : \mathbf{I}^1 \to X$ and $\tau : \mathbf{\Delta}^3 \to X$, satisfying $\sigma(v_0) = x_0$ and $\sigma(v_1) = \tau(v_0)$.

See Figure 3 for an illustration.

Let (Z, z_0) be a pointed polyhedron. By *linear triangular kite* in (Z, z_0) we mean the obvious variant of linear quadrangular kite. Namely this is a pair (σ, τ), consisting of a string σ in Z and a linear map $\tau : \mathbf{\Delta}^3 \to X$. These must satisfy $\sigma(v_0) = z_0$ and $\sigma(v_1) = \tau(v_0)$. Likewise we define *linear triangular balloons.*

Definition 9.1.6. The *boundary* of $\mathbf{\Delta}^3$ is the sequence of linear triangular kites

$$\partial \mathbf{\Delta}^3 = (\partial_1 \mathbf{\Delta}^3, \partial_2 \mathbf{\Delta}^3, \partial_3 \mathbf{\Delta}^3, \partial_4 \mathbf{\Delta}^3)$$

in $(\mathbf{\Delta}^3, v_0)$ defined as follows.

- Let $\sigma_2^\flat, \sigma_3^\flat, \sigma_4^\flat$ be the empty strings in $\mathbf{\Delta}^3$. And let $\sigma_1^\flat : \mathbf{I}^1 \to \mathbf{\Delta}^3$ be the linear map defined on vertices by

$$\sigma_1^\flat(v_0, v_1) := (v_0, v_1).$$

- Let $\tau_i^\flat : \mathbf{\Delta}^2 \to \mathbf{\Delta}^3$ be the linear maps given on vertices by:

$$\tau_1^\flat(v_0, v_1, v_2) := (v_1, v_2, v_3),$$

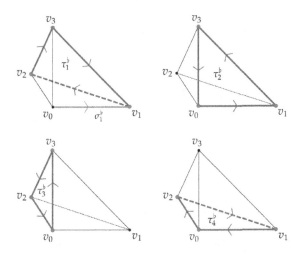

FIGURE 33. The boundary of Δ^3.

$$\tau_2^{\flat}(v_0, v_1, v_2) := (v_0, v_1, v_3),$$

$$\tau_3^{\flat}(v_0, v_1, v_2) := (v_0, v_3, v_2),$$

$$\tau_4^{\flat}(v_0, v_1, v_2) := (v_0, v_2, v_1).$$

• The kite are

$$\partial_i \Delta^3 := (\sigma_i^{\flat}, \tau_i^{\flat}).$$

See Figure 33.

Warning: the kites $(\sigma_i^{\flat}, \tau_i^{\flat})$ above should be confused with the quadrangular kites from Definition 8.1.5, despite the shared notation.

Definition 9.1.7. Let (σ, τ) be a piecewise smooth triangular balloon in (X, x_0). Its *boundary* is the sequence of piecewise smooth triangular kites

$$\partial(\sigma, \tau) := \big(\partial_1(\sigma, \tau), \partial_2(\sigma, \tau), \partial_3(\sigma, \tau), \partial_4(\sigma, \tau)\big)$$

in (X, x_0), where

$$\partial_i(\sigma, \tau) := (\sigma, \tau) \circ (\sigma_i^{\flat}, \tau_i^{\flat}).$$

See Figure 4 for an illustration.

9.2   MI on Triangular Kites

As before, (X, x_0) is a pointed manifold.

Let
$$\mathbf{C} := (G, H, \Psi, \Phi_0)$$
be a Lie quasi crossed module (Definition 5.1.1). As usual we write $\mathfrak{g} := \mathrm{Lie}(G)$ and $\mathfrak{h} := \mathrm{Lie}(H)$. An additive feedback for $\mathbf{C}$ over (X, x_0) is an element
$$\Phi_X \in \mathcal{O}(X) \otimes \mathrm{Hom}(\mathfrak{h}, \mathfrak{g})$$
satisfying condition $(**)$ of Definition 5.2.1. Just like in Definition 5.2.2, we call the data
$$\mathbf{C}/X := (G, H, \Psi, \Phi_0, \Phi_X)$$
a *Lie quasi crossed module with additive feedback* over (X, x_0).

Definition 9.2.1. Let $\alpha \in \Omega^1(X) \otimes \mathfrak{g}$, and let σ be a piecewise smooth path in X. Consider the piecewise smooth differential form
$$\alpha' := \sigma^*(\alpha) \in \Omega^1_{\mathrm{pws}}(\mathbf{I}^1) \otimes \mathfrak{g}.$$
We define
$$\mathrm{MI}(\alpha \mid \sigma) := \mathrm{MI}(\alpha' \mid \mathbf{I}^1) \in G$$
(cf. Definition 3.3.20).

Proposition 9.2.2. *Let $\alpha \in \Omega^1(X) \otimes \mathfrak{g}$.*

(1) *Suppose σ_1 and σ_2 are piecewise smooth paths in X such that $\sigma_1 * \sigma_2$ is defined. Then*
$$\mathrm{MI}(\alpha \mid \sigma_1 * \sigma_2) = \mathrm{MI}(\alpha \mid \sigma_1) \cdot \mathrm{MI}(\alpha \mid \sigma_2).$$

(2) *Let σ be a piecewise smooth path in X. Then*
$$\mathrm{MI}(\alpha \mid \sigma^{-1}) = \mathrm{MI}(\alpha \mid \sigma)^{-1}.$$

(3) *Suppose σ' is a string in a polyhedron Z, $f : Z \to X$ is a piecewise smooth map, and $\sigma := f \circ \sigma'$ is the path in X obtained by the operation (9.1.1). Then*
$$\mathrm{MI}(\alpha \mid \sigma) = \mathrm{MI}(f^*(\alpha) \mid \sigma').$$

Proof. (1) We could use the 2-dimensional Stokes Theorem; but there is an elementary proof. Let $\sigma^1_1, \sigma^1_2 : \mathbf{I}^1 \to \mathbf{I}^1$ be the linear maps belonging to tes[1] $\mathbf{I}^1$, as in Definition 3.1.2. Let $\sigma : \mathbf{I}^1 \to X$ be the piecewise smooth map such that $\sigma \circ \sigma^1_i = \sigma_i$ for $i = 1, 2$. And let $\alpha' := \sigma^*(\alpha)$. Now by definition of $*$ we have $\sigma = \sigma_1 * \sigma_2$ as paths in X. Hence
$$\mathrm{MI}(\alpha \mid \sigma_1 * \sigma_2) = \mathrm{MI}(\alpha \mid \sigma) = \mathrm{MI}(\alpha' \mid \mathbf{I}^1).$$

Likewise
$$\mathrm{MI}(\alpha \,|\, \sigma_i) = \mathrm{MI}(\alpha \,|\, \sigma \circ \sigma_i^1) = \mathrm{MI}(\alpha' \,|\, \sigma_i^1).$$

On the other hand, by Proposition 3.3.22(3) we know that
$$\mathrm{MI}(\alpha' \,|\, \mathbf{I}^1) = \mathrm{MI}(\alpha' \,|\, \sigma_1^1) \cdot \mathrm{MI}(\alpha' \,|\, \sigma_2^1).$$

(2) This is immediate from Proposition 3.5.8(2).

(3) This follows part (1) and induction on the number of pieces in the string σ'. $\qquad\square$

In order to define MI on triangular kites we shall need the following geometric construction. Consider the linear triangular kite (σ', τ') in $(\mathbf{I}^2, v_0)$, where the string σ' has one linear piece $\sigma' : \mathbf{I}^1 \to \mathbf{I}^2$ defined on vertices by
$$\sigma'(v_0, v_1) := (v_0, (\tfrac{1}{2}, \tfrac{1}{2})).$$
The linear map $\tau' : \Delta^2 \to \mathbf{I}^2$ is defined on vertices by
$$\tau'(v_0, v_1, v_2) := ((\tfrac{1}{2}, \tfrac{1}{2}), (1, \tfrac{1}{2}), (\tfrac{1}{2}, 1)).$$
We also need the linear quadrangular kite (σ', τ'') in $(\mathbf{I}^2, v_0)$, where the linear map $\tau'' : \mathbf{I}^2 \to \mathbf{I}^2$ is defined on vertices by
$$\tau''(v_0, v_1, v_2) := ((\tfrac{1}{2}, \tfrac{1}{2}), (1, \tfrac{1}{2}), (\tfrac{1}{2}, 1)).$$
We write $Z := \sigma'(\mathbf{I}^1)$ and $Y := \tau'(\Delta^2)$.

Consider the canonical linear embedding $\Delta^2 \to \mathbf{I}^2$ which on vertices is
$$(v_0, v_1, v_2) \mapsto (v_0, v_1, v_2).$$

Let $h : \mathbf{I}^2 \to \Delta^2$ be the piecewise linear retraction which is linear on Δ^2 and on the triangle complementary to it, and satisfies $h(1, 1) = v_1$. See Figure 34.

Lemma 9.2.3. *There exists a piecewise linear retraction $g : \mathbf{I}^2 \to Y \cup Z$, such that*
$$g \circ \tau'' = \tau' \circ h$$
as piecewise linear maps $\mathbf{I}^2 \to \mathbf{I}^2$.

Proof. Easy exercise. Cf. proof of Proposition 4.1.6. And see Figure 35. $\quad\square$

Let us fix such a retraction g, which we also view as a piecewise linear map $g : \mathbf{I}^2 \to \mathbf{I}^2$.

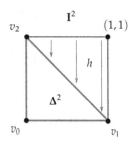

FIGURE 34. The piecewise linear retraction $h : \mathbf{I}^2 \to \Delta^2$.

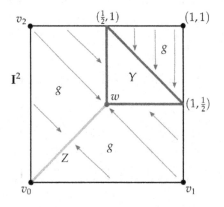

FIGURE 35. The piecewise linear retraction $g : \mathbf{I}^2 \to Y \cup Z$. Here $w := (\tfrac{1}{2}, \tfrac{1}{2})$.

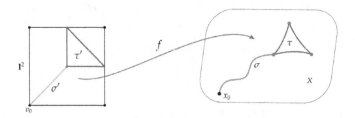

FIGURE 36. The piecewise smooth map $f : \mathbf{I}^2 \to X$ such that $f \circ (\sigma', \tau') = (\sigma, \tau)$.

Definition 9.2.4. Let $\alpha \in \Omega^1(X) \otimes \mathfrak{g}$ and $\beta \in \Omega^1(X) \otimes \mathfrak{h}$. Given a piecewise smooth triangular kite (σ, τ) in (X, x_0) we define its *multiplicative integral*

$$\mathrm{MI}(\alpha, \beta \,|\, \sigma, \tau) \in H$$

as follows. Let $f : \mathbf{I}^2 \to X$ be the unique piecewise smooth map such that:

- $f \circ \sigma' = \sigma$ as maps $\mathbf{I}^1 \to X$.
- $f \circ \tau' = \tau$ as maps $\Delta^2 \to X$.
- $f = f \circ g$ as maps $\mathbf{I}^2 \to X$, where $g : \mathbf{I}^2 \to \mathbf{I}^2$ is the chosen retraction.

(See Figure 36.) We get differential forms

$$\alpha' := f^*(\alpha) \in \Omega^1_{\mathrm{pws}}(\mathbf{I}^2) \otimes \mathfrak{g}$$

and

$$\beta' := f^*(\beta) \in \Omega^2_{\mathrm{pws}}(\mathbf{I}^2) \otimes \mathfrak{h}.$$

We define

$$\mathrm{MI}(\alpha, \beta \,|\, \sigma, \tau) := \mathrm{MI}(\alpha', \beta' \,|\, \sigma', \tau''),$$

where (σ', τ'') is the linear quadrangular kite in $(\mathbf{I}^2, v_0)$ defined above, and $\mathrm{MI}(\alpha', \beta' \,|\, \sigma', \tau'')$ is the multiplicative integral from Definition 4.4.16.

This definition might seem strange; but we shall soon see that it has all the expected good properties.

Proposition 9.2.5 (Functoriality in X). *Let $e : (Y, y_0) \to (X, x_0)$ be a map of pointed manifolds, let $\alpha \in \Omega^1(X) \otimes \mathfrak{g}$, let $\beta \in \Omega^2(X) \otimes \mathfrak{h}$, and let (σ, τ) be a piecewise smooth triangular kite in (Y, y_0). Then*

$$\mathrm{MI}(\alpha, \beta \,|\, e \circ \sigma, e \circ \tau) = \mathrm{MI}(e^*(\alpha), e^*(\beta) \,|\, \sigma, \tau),$$

where the latter is calculated with respect to the Lie quasi crossed module with additive feedback $e^*(\mathbf{C}/X)$.

Proof. This is immediate from the definition, since $(e \circ f)^* = f^* \circ e^*$ for a piecewise smooth map $f : \mathbf{I}^2 \to Y$. $\square$

Proposition 9.2.6 (Comparison to Quadrangular Kites). *Let (Z, z_0) be a pointed polyhedron, $e : (Z, z_0) \to (X, x_0)$ a piecewise smooth map preserving base points, (σ, τ) a piecewise smooth triangular kite in (X, x_0), (σ', τ') a linear quadrangular kite in (Z, z_0), $\alpha \in \Omega^1(X) \otimes \mathfrak{g}$ and $\beta \in \Omega^2(X) \otimes \mathfrak{h}$. Assume that $\sigma = e \circ \sigma'$ as paths in X, and $\tau \circ h = e \circ \tau'$ as maps $\mathbf{I}^2 \to X$, where $h : \mathbf{I}^2 \to \Delta^2$ is the retraction in Figure 34. Then*

$$\mathrm{MI}(\alpha, \beta \,|\, \sigma, \tau) = \mathrm{MI}(e^*(\alpha), e^*(\beta) \,|\, \sigma', \tau'),$$

where the latter is calculated with respect to the Lie quasi crossed module with additive feedback $e^*(\mathbf{C}/X)$.

Proof. The piecewise smooth map $f : \mathbf{I}^2 \to X$ used in Definition 9.2.4 can be factored as $f = e \circ f'$ for a piecewise linear map $f' : (\mathbf{I}^2, v_0) \to (Z, z_0)$ of pointed polyhedra. The assertion now follows from Proposition 4.5.3.   □

The notion of compatible connection from Definition 5.2.4 makes sense here too, only we have to consider $\alpha \in \Omega^1(X) \otimes \mathfrak{g}$ (and replace "string" with "piecewise smooth path").

Proposition 9.2.7 (Moving the Base Point). *Let* $\alpha \in \Omega^1(X) \otimes \mathfrak{g}$ *be a connection compatible with* $\mathbf{C}/X$, *and let* ρ *be a piecewise smooth path in* X, *with initial point* x_0 *and terminal point* x_1. *Define* $g := \mathrm{MI}(\alpha \,|\, \rho)$, *and let*

$$\mathbf{C}^{\mathrm{g}}/X = (G, H^{\mathrm{g}}, \Psi, \Phi_0^{\mathrm{g}}, \Phi_X)$$

be the Lie quasi crossed module with additive feedback over (X, x_1) *constructed in Section 5.4. Given a form* $\beta \in \Omega^2(X) \otimes \mathfrak{h}$ *and a piecewise smooth triangular kite* (σ, τ) *in* (X, x_1), *consider the element*

$$\mathrm{MI}^{\mathrm{g}}(\alpha, \beta \,|\, \sigma, \tau) \in H^{\mathrm{g}}$$

calculated with respect to the Lie quasi crossed module with additive feedback $\mathbf{C}^{\mathrm{g}}/X$. *Then*

$$\Psi(g)\big(\mathrm{MI}^{\mathrm{g}}(\alpha, \beta \,|\, \sigma, \tau)\big) = \mathrm{MI}(\alpha, \beta \,|\, \rho * \sigma, \tau)$$

in H.

Proof. It is possible to find a polyhedron Z, points $z_0, z_1 \in Z$, a string ρ' in Z with initial point z_0 and terminal point z_1, a linear quadrangular kite (σ', τ') in (Z, z_1) and a piecewise smooth map $e : Z \to X$, such that $\rho = e \circ \rho'$ as paths and $(\sigma, \tau) = e \circ (\sigma', \tau')$ as kites. Define $\alpha' := e^*(\alpha)$ and $\beta' := e^*(\beta)$.

Let $\mathbf{C}'/Z := e^*(\mathbf{C}/X)$ be the induced Lie quasi crossed module with additive feedback over the pointed polyhedron (Z, z_0). By Proposition 9.2.2(3) we know that $\mathrm{MI}(\alpha' \,|\, \rho') = g$. Examining the definition of $\mathbf{C}^{\mathrm{g}}/X$ in Section 5.4 we see that

$$e^*(\mathbf{C}^{\mathrm{g}}/X) = (\mathbf{C}')^{\mathrm{g}}/Z$$

as Lie quasi crossed modules with additive feedback over the pointed polyhedron (Z, z_1). Therefore $\mathrm{MI}^{\mathrm{g}}(\alpha', \beta' \,|\, \sigma', \tau')$ is unambiguous.

By Proposition 9.2.6 we know that

$$\mathrm{MI}^{\mathrm{g}}(\alpha', \beta' \,|\, \sigma', \tau') = \mathrm{MI}^{\mathrm{g}}(\alpha, \beta \,|\, \sigma, \tau)$$

and

$$\mathrm{MI}(\alpha', \beta' \,|\, \rho' * \sigma', \tau') = \mathrm{MI}(\alpha, \beta \,|\, \rho * \sigma, \tau).$$

Finally, by Theorem 5.4.8 we have

$$\mathrm{MI}^g(\alpha', \beta' \,|\, \sigma', \tau') = \mathrm{MI}(\alpha', \beta' \,|\, \rho' * \sigma', \tau').$$

$\square$

9.3 Stokes Theorems

We continue with the setup from before: (X, x_0) is a pointed manifold, and

$$\mathbf{C}/X = (G, H, \Psi, \Phi_0, \Phi_X)$$

be a Lie quasi crossed module with additive feedback.

Just like in Definition 5.3.1, a *connection-curvature pair* for $\mathbf{C}/X$ is a pair (α, β), with $\alpha \in \Omega^1(X) \otimes \mathfrak{g}$ and $\beta \in \Omega^1(X) \otimes \mathfrak{h}$, such that α is a compatible connection, and the differential equation

$$\Phi_X(\beta) = \mathrm{d}(\alpha) + \tfrac{1}{2}[\alpha, \alpha]$$

holds in $\Omega^2(X) \otimes \mathfrak{g}$.

Recall that for a piecewise smooth triangular kite (σ, τ), its boundary $\partial(\sigma, \tau)$ was defined in Definition 9.1.4.

Theorem 9.3.1 (Stokes Theorem for Triangles). *Let (X, x_0) be a pointed manifold, let*

$$\mathbf{C}/X := (G, H, \Psi, \Phi_0, \Phi_X)$$

be a Lie quasi crossed module with additive feedback over (X, x_0), let (α, β) be a connection-curvature pair for $\mathbf{C}/X$, and let (σ, τ) be a piecewise smooth triangular kite in (X, x_0). Then

$$\Phi_0\big(\mathrm{MI}(\alpha, \beta \,|\, \sigma, \tau)\big) = \mathrm{MI}(\alpha, \beta \,|\, \partial(\sigma, \tau)).$$

Proof. As in the proof of Proposition 9.2.7, we can find a pointed polyhedron (Z, z_0), a linear quadrangular kite (σ', τ') in (Z, z_0) and a piecewise smooth map $f : Z \to X$, such that $(\sigma, \tau) = f \circ (\sigma', \tau')$ as kites. Define $\alpha' := f^*(\alpha)$ and $\beta' := f^*(\beta)$. According to Proposition 9.2.2(3) we have

$$\mathrm{MI}(\alpha, \beta \,|\, \partial(\sigma, \tau)) = \mathrm{MI}(\alpha' \,|\, \partial(\sigma', \tau')).$$

And by Proposition 9.2.6 we have

$$\mathrm{MI}(\alpha, \beta \,|\, \sigma, \tau) = \mathrm{MI}(\alpha', \beta' \,|\, \sigma', \tau').$$

Finally by Theorem 6.2.1 we know that

$$\mathrm{MI}(\alpha' \,|\, \partial(\sigma', \tau')) = \Phi_0\big(\mathrm{MI}(\alpha', \beta' \,|\, \sigma', \tau')\big).$$

$\square$

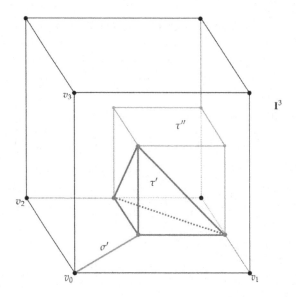

FIGURE 37. The triangular linear balloon (σ', τ') and the quadrangular linear balloon (σ', τ'') in the pointed polyhedron $(\mathbf{I}^3, v_0)$.

For the 3-dimensional Stokes Theorem we shall need an auxiliary geometric construction, similar to the one in the previous section. Consider the linear triangular balloon (σ', τ') in $(\mathbf{I}^3, v_0)$, where the string σ' has one linear piece $\sigma' : \mathbf{I}^1 \to \mathbf{I}^3$ defined on vertices by

$$\sigma'(v_0, v_1) := \left(v_0, (\tfrac{1}{2}, \tfrac{1}{2}, 0)\right).$$

The linear map $\tau' : \mathbf{\Delta}^3 \to \mathbf{I}^3$ is defined on vertices by

$$\tau'(v_0, v_1, v_2, v_3) := \left((\tfrac{1}{2}, \tfrac{1}{2}, 0), (1, \tfrac{1}{2}, 0), (\tfrac{1}{2}, 1, 0), (\tfrac{1}{2}, \tfrac{1}{2}, \tfrac{1}{2})\right).$$

We also need the linear quadrangular balloon (σ', τ'') in $(\mathbf{I}^3, v_0)$, where the linear map $\tau'' : \mathbf{I}^3 \to \mathbf{I}^3$ is defined on vertices by the same formula as τ'. We let $Z := \sigma'(\mathbf{I}^1)$ and $Y := \tau'(\mathbf{\Delta}^3)$. See Figure 37.

Consider the canonical linear embedding $\mathbf{\Delta}^3 \to \mathbf{I}^3$ which on vertices is

$$(v_0, \ldots, v_3) \mapsto (v_0, \ldots, v_3).$$

Let $h : \mathbf{I}^3 \to \mathbf{\Delta}^3$ be the piecewise linear retraction $h := h_2 \circ h_1$, where h_1 and h_2 are the retractions shown in Figure 38.

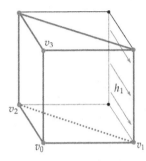 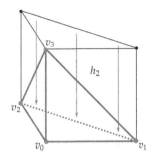

FIGURE 38. The piecewise linear map h_1 retracts the cube $\mathbf{I}^3$ to a prism. The piecewise linear map h_2 retracts the prism to the tetrahedron Δ^3.

Lemma 9.3.2. *There is a piecewise linear retraction* $g : \mathbf{I}^3 \to Z \cup Y$, *such that*

$$g \circ \tau'' = \tau' \circ h$$

as maps $\mathbf{I}^3 \to \mathbf{I}^3$.

Proof. Nice exercise in piecewise linear geometry. $\qquad\square$

Let us fix such a retraction g, which we also view as a piecewise linear map $g : \mathbf{I}^3 \to \mathbf{I}^3$.

The notions of *tame connection* and *inert differential form*, from Sections 5.3 and 8.3 respectively, make sense here, and all pertinent results hold. Given a tame connection $\alpha \in \Omega^1(X) \otimes \mathfrak{g}$, an inert form $\gamma \in \Omega^3(X) \otimes \mathfrak{h}$ and a piecewise smooth balloon (σ, τ) in (X, x_0), we define the *twisted multiplicative integral*

$$\mathrm{MI}(\alpha, \gamma \,|\, \sigma, \tau) \in H_0$$

as follows. Let $f : \mathbf{I}^3 \to X$ be the unique piecewise smooth map satisfying:

- $f \circ \sigma' = \sigma$ as maps $\mathbf{I}^1 \to X$.
- $f \circ \tau' = \tau$ as maps $\Delta^3 \to X$.
- $f = f \circ g$ as maps $\mathbf{I}^3 \to X$, where $g : \mathbf{I}^3 \to \mathbf{I}^3$ is the chosen retraction.

Now

$$\mathbf{C}'/\mathbf{I}^3 := f^*(\mathbf{C}/X) = (G, H, \Psi, \Phi_0, f^*(\Phi_X))$$

is a Lie quasi crossed module with additive feedback over $(\mathbf{I}^3, v_0)$,

$$\alpha' := f^*(\alpha) \in \Omega^1_{\mathrm{pws}}(\mathbf{I}^3) \otimes \mathfrak{g}$$

is a tame connection for $\mathbf{C}'/X$, and

$$\gamma' := f^*(\gamma) \in \Omega^3_{\mathrm{pws}}(\mathbf{I}^3) \otimes \mathfrak{h}$$

is an inert differential form. We define

(9.3.3) $\mathrm{MI}(\alpha, \gamma \mid \sigma, \tau) := \mathrm{MI}(\alpha', \gamma' \mid \tau''),$

where $\tau'' : \mathbf{I}^3 \to \mathbf{I}^3$ is the linear map described above, and $\mathrm{MI}(\alpha', \gamma' \mid \tau'')$ is the twisted multiplicative integral from Definition 8.3.12.

Remark 9.3.4. The reason that the path σ is explicit in the expression $\mathrm{MI}(\alpha, \gamma \mid \sigma, \tau)$ is that this operation could depend on the homotopy class of σ, in case the manifold X is not simply connected. Since polyhedra are always simply connected, we did not have to worry about strings in Definition 8.3.12.

It is easy to see that $\mathrm{MI}(\alpha, \gamma \mid \sigma, \tau)$ makes sense for an inert p-form γ and a "p-dimensional balloon" (σ, τ), for any $p \geq 0$. Moreover, for $p = 2$ this coincides with the nonabelian MI of Definition 9.2.4.

Recall the boundary $\partial(\sigma, \tau)$ of a balloon (σ, τ) in (X, x_0), from Definition 9.1.7. As is our convention for the multiplicative integral of a sequence, we write

$$\mathrm{MI}(\alpha, \beta \mid \partial(\sigma, \tau)) := \prod_{i=1,\dots,4} \mathrm{MI}(\alpha, \beta \mid \partial_i(\sigma, \tau)).$$

As in Definition 8.6.1, the 3-curvature of (α, β) is the form

$$\gamma := \mathrm{d}(\beta) + \psi_{\mathfrak{h}}(\alpha)(\beta) \in \Omega^3(X) \otimes \mathfrak{h}.$$

Here is the second version of the main result of the book (the first version was Theorem 8.6.6).

Theorem 9.3.5 (Stokes Theorem for Tetrahedra). *Let (X, x_0) be a pointed manifold, let $\mathbf{C}/X$ be a Lie quasi crossed module with additive feedback, let (α, β) be a connection-curvature pair for $\mathbf{C}/X$, and let γ be the 3-curvature of (α, β). Then:*

(1) *The differential form γ is inert.*
(2) *For any piecewise smooth balloon (σ, τ) in (X, x_0) one has*

$$\mathrm{MI}(\alpha, \beta \mid \partial(\sigma, \tau)) = \mathrm{MI}(\alpha, \gamma \mid \sigma, \tau)$$

in H.

Proof. (1) It suffices to prove that for any piecewise smooth map $f : \mathbf{I}^3 \to X$ the form

$$\gamma' := f^*(\gamma) \in \Omega^3_{\mathrm{pws}}(\mathbf{I}^3) \otimes \mathfrak{h}$$

is inert. Define
$$\alpha' := f^*(\alpha) \in \Omega^1_{\mathrm{pws}}(\mathbf{I}^3) \otimes \mathfrak{g}$$
and
$$\beta' := f^*(\beta) \in \Omega^2_{\mathrm{pws}}(\mathbf{I}^3) \otimes \mathfrak{h}.$$

Then (α', β') is a connection curvature pair in $\mathbf{C}'/\mathbf{I}^3$ (this is a variant of Proposition 5.3.3(1)), and γ' is the 3-curvature of (α', β'). But according to Theorem 8.6.6(1) the form γ' is inert.

(2) Let $f : (\mathbf{I}^3, v_0) \to (X, x_0)$ be the pointed piecewise smooth map constructed just before (9.3.3). By definition we have
$$\mathrm{MI}(\alpha, \gamma \,|\, \sigma, \tau) := \mathrm{MI}(\alpha', \gamma' \,|\, \tau'').$$

By Proposition 9.2.6 we know that
$$\mathrm{MI}\big(\alpha, \beta \,|\, \partial(\sigma, \tau)\big) = \mathrm{MI}\big(\alpha', \beta' \,|\, \partial(\sigma', \tau')\big).$$

And by Theorem 8.6.6(2) we know that
$$\mathrm{MI}\big(\alpha', \beta' \,|\, \partial(\sigma', \tau'')\big) = \mathrm{MI}\big(\alpha', \gamma' \,|\, \tau''\big).$$

It remains to prove that

(9.3.6) $\qquad \mathrm{MI}\big(\alpha', \beta' \,|\, \partial(\sigma', \tau')\big) = \mathrm{MI}\big(\alpha', \beta' \,|\, \partial(\sigma', \tau'')\big).$

By definition we have
$$\mathrm{MI}\big(\alpha', \beta' \,|\, \partial(\sigma', \tau')\big) = \prod_{i=1,\ldots,4} \mathrm{MI}\big(\alpha, \beta \,|\, \partial_i(\sigma', \tau')\big)$$
and
$$\mathrm{MI}\big(\alpha', \beta' \,|\, \partial(\sigma', \tau'')\big) = \prod_{i=1,\ldots,6} \mathrm{MI}\big(\alpha, \beta \,|\, \partial_i(\sigma', \tau'')\big).$$

Using Proposition 9.2.6 and looking at Figures 37, 33 and 31 we see that
$$\mathrm{MI}\big(\alpha', \beta' \,|\, \partial_1(\sigma', \tau'')\big) = 1,$$
$$\mathrm{MI}\big(\alpha', \beta' \,|\, \partial_2(\sigma', \tau'')\big) = 1,$$
$$\mathrm{MI}\big(\alpha', \beta' \,|\, \partial_3(\sigma', \tau'')\big) = \mathrm{MI}\big(\alpha', \beta' \,|\, \partial_1(\sigma', \tau')\big),$$
$$\mathrm{MI}\big(\alpha', \beta' \,|\, \partial_4(\sigma', \tau'')\big) = \mathrm{MI}\big(\alpha', \beta' \,|\, \partial_2(\sigma', \tau')\big),$$
$$\mathrm{MI}\big(\alpha', \beta' \,|\, \partial_5(\sigma', \tau'')\big) = \mathrm{MI}\big(\alpha', \beta' \,|\, \partial_3(\sigma', \tau')\big)$$
and
$$\mathrm{MI}\big(\alpha', \beta' \,|\, \partial_6(\sigma', \tau'')\big) = \mathrm{MI}\big(\alpha', \beta' \,|\, \partial_4(\sigma', \tau')\big).$$

Thus equation (9.3.6) is true. $\qquad\qquad\qquad\qquad\qquad\qquad\qquad\qquad\square$

9.4   Rationality of the Multiplicative Integral

Let $\mathbb{K}$ be a subfield of $\mathbb{R}$. As common in algebraic geometry, let us denote by $\mathbf{A}^n(\mathbb{K})$ the set of $\mathbb{K}$-rational points of $\mathbf{A}^n(\mathbb{R})$; namely those points $x \in \mathbf{A}^n(\mathbb{R})$ whose coordinates satisfy $t_i(x) \in \mathbb{K}$, $i = 1, \ldots, n$.

By (embedded) *polyhedron defined over* $\mathbb{K}$ we mean a polyhedron $X \subset \mathbf{A}^n(\mathbb{R})$, such that all the vertices of X belong to $\mathbf{A}^n(\mathbb{K})$. For such X, and for every field L such that $\mathbb{K} \subset L \subset \mathbb{R}$, we can talk about the set $X(L)$ of L-rational points of X. The real polyhedron is of course $X(\mathbb{R})$.

Suppose Y is another polyhedron defined over $\mathbb{K}$. It makes sense to talk about linear maps $f : X \to Y$ defined over $\mathbb{K}$. The condition is of course that for any vertex $v \in X$ the point $f(v)$ is in $Y(\mathbb{K})$.

The simplices Δ^p are defined over $\mathbb{Q}$. Therefore we have the notion of linear triangulation of X defined over $\mathbb{K}$. This lets us consider piecewise linear maps $f : Y \to X$ defined over $\mathbb{K}$. In particular we have *kites in X defined over* $\mathbb{K}$.

The polyhedron X has a $\mathbb{K}$-subalgebra $\mathcal{O}_{\mathrm{alg}}(X(\mathbb{K})) \subset \mathcal{O}(X)$ of $\mathbb{K}$-*valued algebraic functions on* $X(\mathbb{K})$, which is the restriction to $X(\mathbb{K})$ of the polynomial algebra $\mathbb{K}[t_1, \ldots, t_n] \subset \mathcal{O}(\mathbf{A}^n(\mathbb{R}))$. Similarly one can define the DG algebra $\Omega_{\mathrm{alg}}(X(\mathbb{K}))$ of $\mathbb{K}$-*valued algebraic differential forms*, which is contained in $\Omega(X)$. Using linear triangulations defined over $\mathbb{K}$ we can also consider the $\mathbb{K}$-algebra $\mathcal{O}_{\mathrm{pwa}}(X(\mathbb{K}))$ of $\mathbb{K}$-*valued piecewise algebraic functions*, and the DG algebra $\Omega_{\mathrm{pwa}}(X(\mathbb{K}))$ of $\mathbb{K}$-*valued piecewise algebraic differential forms*, which is a DG subalgebra of $\Omega_{\mathrm{pws}}(X)$.

Suppose X is a polyhedron and $Y \subset \mathbf{A}^m$ is an affine algebraic variety, both defined over $\mathbb{K}$. A map $f : X \to Y$ is said to be algebraic and defined over $\mathbb{K}$ if for every i the function $t_i \circ f : X(\mathbb{K}) \to \mathbb{R}$ belongs to $\mathcal{O}_{\mathrm{alg}}(X(\mathbb{K}))$. Equivalently, f extends to a map of algebraic varieties $\tilde{f} : \mathbf{A}^n \to \mathbf{A}^m$ that is defined over $\mathbb{K}$. Using linear triangulations defined over $\mathbb{K}$ we can also consider piecewise algebraic maps $f : X \to Y$ defined over $\mathbb{K}$.

Now assume that we are given a Lie quasi crossed module with additive feedback

$$\mathbf{C} = (G, H, \Psi, \Phi_0, \Phi_X)$$

in which (X, x_0) is a pointed polyhedron defined over $\mathbb{K}$, the groups G and H are *affine unipotent* linear algebraic groups defined over $\mathbb{K}$, the maps $\Psi : G \times H \to H$ and $\Phi_0 : H \to G$ are maps of algebraic varieties defined over $\mathbb{K}$, and

$$\Phi_X \in \mathcal{O}_{\mathrm{pwa}}(X(\mathbb{K})) \otimes_{\mathbb{K}} \mathrm{Hom}_{\mathbb{K}}(\mathfrak{h}(\mathbb{K}), \mathfrak{g}(\mathbb{K})).$$

We say that $\mathbf{C}/X$ is an *algebraic unipotent quasi crossed module with additive feedback defined over* $\mathbb{K}$. Note that the Lie groups in this setup are $G(\mathbb{R})$ and $H(\mathbb{R})$. Also note that exponential maps $\exp_G : \mathfrak{g} \to G$ and $\exp_H : \mathfrak{h} \to H$ are isomorphisms of algebraic varieties defined over $\mathbb{K}$ – this is because the Lie algebras $\mathfrak{h}$ and $\mathfrak{g}$ are nilpotent.

The polyhedron $\mathbf{I}^1$ is defined over $\mathbb{Q}$. Suppose we are given a piecewise algebraic differential form

$$\alpha \in \Omega^p_{\mathrm{pwa}}(\mathbf{I}^1(\mathbb{K})) \otimes_{\mathbb{K}} \mathfrak{g}(\mathbb{K}).$$

It is not hard to show (using the ODE (5.5.2), and the fact that $\exp_G$ is algebraic) that the map $g : \mathbf{I}^1 \to G$ of (5.5.1) is piecewise algebraic and defined over $\mathbb{K}$. In particular we get

$$g(1) = \mathrm{MI}(\alpha \,|\, \mathbf{I}^1) \in G(\mathbb{K}).$$

From this it follows that for any polyhedron X defined over $\mathbb{K}$, any

$$\alpha \in \Omega^1_{\mathrm{alg}}(X(\mathbb{K})) \otimes_{\mathbb{K}} \mathfrak{g}(\mathbb{K})$$

and any piecewise algebraic string σ in X defined over $\mathbb{K}$, one has

$$\mathrm{MI}(\alpha \,|\, \sigma) \in G(\mathbb{K}).$$

We are led to make the following conjecture. A connection-curvature pairs (α, β) is said to be algebraic if $\alpha \in \Omega^1_{\mathrm{alg}}(X(\mathbb{K})) \otimes_{\mathbb{K}} \mathfrak{g}(\mathbb{K})$ and $\alpha \in \Omega^2_{\mathrm{alg}}(X(\mathbb{K})) \otimes_{\mathbb{K}} \mathfrak{h}(\mathbb{K})$.

Conjecture 9.4.1. Let $\mathbf{C}/X$ be an algebraic unipotent quasi crossed module with additive feedback defined over $\mathbb{K}$, let (α, β) be an algebraic connection-curvature pair in $\mathbf{C}/X$, and let (σ, τ) be a piecewise algebraic kite in (X, x_0). Then

$$\mathrm{MI}(\alpha, \beta \,|\, \sigma, \tau) \in H(\mathbb{K}) .$$

It seems that the very recent paper [BGNT] may have a proof of this conjecture in the special case when $\mathbf{C}/X$ is a *crossed module* and *the 3-curvature of (α, β) vanishes*. See Section 5.5.

9.5   A Conjecture on Descent Data

Suppose $\mathfrak{f}$ is a cosimplicial nilpotent quantum type DG Lie algebra, just like in Section 5.7. Take an MC element $\tilde{\omega} = \{\tilde{\omega}^{p,q}\}$ in the Thom-Sullivan normalization $\tilde{\mathrm{N}}(\mathfrak{f})$. Then

$$\tilde{\omega}^{0,0} \in \mathcal{O}(\Delta^0) \otimes \mathfrak{f}^{0,1} = \mathfrak{f}^{0,1},$$

$$\tilde{\omega}^{1,1} \in \Omega^1(\Delta^1) \otimes \mathfrak{f}^{1,0}$$

and
$$\tilde{\omega}^{2,2} \in \Omega^2(\Delta^2) \otimes \mathfrak{f}^{2,-1}.$$

Let us write
$$\omega := \tilde{\omega}^{0,0} \in \mathfrak{f}^{0,1}$$

(this notation is different from the one on Section 5.7),
$$g := \mathrm{MI}(\tilde{\omega}^{1,1} \,|\, \Delta^1) \in \exp(\mathfrak{f}^{1,0})$$

and
$$a := \mathrm{MI}(\tilde{\omega}^{2,1}, \tilde{\omega}^{2,2} \,|\, \Delta^2) \in \exp(\mathfrak{f}^{2,-1})_{\tilde{\omega}^{2,0}(v_0)}.$$

This last multiplicative integral takes place in the Lie quasi crossed module with additive feedback
$$\mathbf{C}_{\tilde{\omega}^{2,0}(v_0)}/\Delta^2 = \big(G, H_{\tilde{\omega}^{2,0}(v_0)}, \Psi, \Phi_{\tilde{\omega}^{2,0}(v_0)}, \Phi_{\Delta^2}\big)$$

as in (5.7.1). In this way we assign to each $\tilde{\omega} \in \mathrm{MC}(\tilde{\mathrm{N}}(\mathfrak{f}))$ a triple

(9.5.1)
$$\mathrm{MI}(\tilde{\omega}\,|\,\Delta) := (\omega, g, a).$$

On the other hand, for every p the nilpotent quantum type DG Lie algebra $\mathfrak{f}^{p,\cdot}$ has its *Deligne 2-groupoid* $\mathrm{Del}(\mathfrak{f}^{p,\cdot})$, also known as the Deligne crossed groupoid; see [Ye5, Section 6]. The collection $\{\mathrm{Del}(\mathfrak{f}^{p,\cdot})\}_{p\in\mathbb{N}}$ is then a cosimplicial crossed groupoid. As explained in [Ye3, Section 5], there is the set of descent data $\mathrm{Desc}(\mathrm{Del}(\mathfrak{f}))$, and its quotient set $\overline{\mathrm{Desc}}(\mathrm{Del}(\mathfrak{f}))$.

Consider an MC element $\tilde{\omega} \in \mathrm{MC}(\tilde{\mathrm{N}}(\mathfrak{f}))$, and the triple $\mathrm{MI}(\tilde{\omega}\,|\,\Delta) = (\omega, g, a)$ from (9.5.1). The 2-dimensional Stokes Theorem (Theorem 9.3.1) implies that (ω, g, a) satisfies the "failure of 1-cocycle" condition of [Ye3, Definition 5.2]. The 3-dimensional Stokes Theorem (Theorem 9.3.5) implies that (ω, g, a) satisfies the "twisted 2-cocycle" condition of loc. cit. This means that $(\omega, g, a) \in \mathrm{Desc}(\mathrm{Del}(\mathfrak{f}))$.

Recall that the quotient set of $\mathrm{MC}(\tilde{\mathrm{N}}(\mathfrak{f}))$ by the action of the gauge group $\exp(\tilde{\mathrm{N}}(\mathfrak{f})^0)$ is denoted by $\overline{\mathrm{MC}}(\tilde{\mathrm{N}}(\mathfrak{f}))$.

Conjecture 9.5.2. Let $\mathfrak{f}$ be a cosimplicial nilpotent quantum type DG Lie algebra. Then the function
$$\mathrm{MI}(-\,|\,\Delta) : \mathrm{MC}(\tilde{\mathrm{N}}(\mathfrak{f})) \to \mathrm{Desc}(\mathrm{Del}(\mathfrak{f}))$$

induces a bijection
$$\overline{\mathrm{MC}}(\tilde{\mathrm{N}}(\mathfrak{f})) \to \overline{\mathrm{Desc}}(\mathrm{Del}(\mathfrak{f})).$$

Bibliography

[BGNT] P. Bressler, A. Gorokhovsky, R. Nest and B. Tsygan, Formality for Algebroids I: Nerves of Two-Groupoids, eprint arXiv:1211.6603.

[Bo] N. Bourbaki, "Lie Groups and Lie Algebras", Chapters 1-3, Springer, 1989.

[BM] L. Breen and W. Messing, Differential geometry of gerbes, *Advances Math.* **198**, Issue 2 (2005), 732-846.

[BS] J.C. Baez and U. Schreiber, Higher gauge theory, in "Categories in Algebra, Geometry and Mathematical Physics", *Contemporary Mathematics* **431** (2007), AMS.

[DF] J. D. Dollard and C. N. Friedman, "Product integration with applications to differential equations", Encyclopedia of mathematics and its applications, Volume 10, Addison-Wesley, 1979.

[FHT] Y. Felix, S. Helperin and J.-C. Thomas, "Rational Homotopy Theory", Springer, 2001.

[Ge] E. Getzler, A Darboux theorem for Hamiltonian operators in the formal calculus of variations, *Duke Math. J.* **111**, Number 3 (2002), 535-560.

[Ha] R. Hartshorne, "Algebraic Geometry", Springer-Verlag, New-York, 1977.

[Jo] D. Joyce, On manifolds with corners, eprint arXiv:0910.3518.

[KMR] R.L. Karp, F. Mansouri and J.S. Rno, Product integral formalism and non-Abelian Stokes theorem, *J. Math. Phys.* **40** (1999), 6033-6043.

[Ko] M. Kontsevich, Deformation quantization of algebraic varieties, *Lett. Math. Phys.* **56** (2001), no. 3, 271-294.

[Le] J.M. Lee, "Introduction to Smooth Manifolds", Springer.

[Pr] M. Prasma, Higher descent data as a homotopy limit, *Journal of Homotopy and Related Structures* (2013), DOI 10.1007/s40062-013-0048-1, eprint arXiv:1112.3072.

[SW1] U. Schreiber and K. Waldorf, Parallel transport and functors, *J. Homotopy Relat. Struct.* **4** (2009), 187-244.

[SW2] U. Schreiber and K. Waldorf, Smooth functors vs. differential forms, *Homology, Homotopy and Applications* **13.1** (2011), 143-203.

[Va] V.L. Varadarajan, "Lie Groups, Lie Algebras, and Their Representations", Springer, 1984.

[Wa] F.W. Warner, "Foundations of Differentiable Manifolds and Lie Groups", Springer, 1983.

[Ye1] A. Yekutieli, Deformation Quantization in Algebraic Geometry, *Advances Math.* **198** (2005), 383-432. Erratum: *Advances Math.* **217** (2008), 2897-2906.

[Ye2]   A. Yekutieli, Mixed Resolutions and Simplicial Sections, *Israel J. Math.* **162** (2007), 1-27.

[Ye3]   A. Yekutieli, Twisted Deformation Quantization of Algebraic Varieties, *Advances in Mathamatics* **268** (2015), 241-305.

[Ye4]   A. Yekutieli, Twisted Deformation Quantization of Algebraic Varieties (Survey), in "New Trends in Noncommutative Algebra", *Contemp. Math.* **562** (2012), pages 279-297.

[Ye5]   A. Yekutieli, MC Elements in Pronilpotent DG Lie Algebras, *J. Pure Appl. Algebra* **216** (2012), 2338-2360.

[Ye6]   A. Yekutieli, Combinatorial Descent Data for Gerbes, to appear in *J. Noncommutative Geometry*, eprint arXiv:1109.1919.

Index